Fundamentals of Agronomy

About the Author

Dr. Sharanappa obtained his B.Sc.(Agriculture), M.Sc. (Agriculture) and Ph.D. (Agronomy) from University of Agricultural Sciences Bangalore. He served in the University of Agricultural Sciences as Assistant Professor Associate Professor, Professor and Professor and Head of the Department of Agronomy for 34 years in teaching, research and extension. At present as Emeritus Professor of ICAR. Under his guidance nine students secured PhD. (Agronomy) and 14 students M. Sc.(Agriculture) in the University of Agricultural Sciences, Bangalore. He served as member of the advisory committee of 68 postgraduate students. His special interest in teaching and research are Organic Farming, Soil Fertility and Nutrient Management in Crops and Cropping Systems. He has developed technologies for nutrient management for organic groundnut, onion, chilli, baby corn and finger millet production. He served as the Principal Investigator for DST funded research project on establishment of organic farming technology park in agroclimatalogically disadvantaged region of Karnataka state and NPOF funded project on establishment of Model Organic Farm at GKVK, Bengaluru. Thirty three research papers were published by him in NAAS rated scientific journals, and published the book on Oilseed Production. He has presented research papers in the international conferences held in Israel and USA on the precision nutrient management in maize and groundnut.

Fundamentals of Agronomy

Sharanappa
Emeritus Professor
Department of Agronomy
College of Agriculture
University of Agricultural Sciences
GKVK, Bengaluru, Karnataka

NEW INDIA PUBLISHING AGENCY
New Delhi-110 034

NEW INDIA PUBLISHING AGENCY
101, Vikas Surya Plaza, CU Block, LSC Market
Pitam Pura, New Delhi – 110 034, India
Email: info@nipabooks.com
Web: www.nipabooks.com

For customer assistance, please contact
Phone: +91-11-27 34 17 17 Fax:+91-11-27 34 16 16
E-Mail: feedbacks@nipabooks.com

ISBN : 978-93-90591-15-2

Composed and Designed by NIPA.

Preface

Agronomy is the core subject of agricultural sciences. A strong foundation for agricultural education emerges by understanding the principles of agronomy. Majority of the students who join for the degree programmes in agriculture universities are never exposed to the crops and crop production aspects. The beginners to agriculture needs to be introduced the meaning, scope and principles of agronomy. With this background and keeping in mind the common syllabus prescribed by the Dean's committee for agriculture education in the country, efforts are made to compile and present the information on introduction to agronomy, agriculture, meaning, scope and importance, history of development of agriculture, agricultural education and research in the country and elsewhere, factors affecting plant growth, manures and fertilizers, principles of agronomy *viz.* soil management, crop management, weed management, management of problematic soils, cropping systems, selection of planting materials, their treatment in simple language. At the end of each chapter subjective and objective questions are provided for practice. The author hopes that this book will be useful for the students for learning and for the faculty for teaching the course "Fundamentals of Agronomy" for the undergraduate degree programme in the Agricultural Universities.

Author

Contents

Chapter 1

Agriculture and Agronomy: History and Development

The term agriculture is derived from the Latin words **ager** or **agri** meaning soils and **cultura** means cultivation. Agriculture is the art, science and business of crop production. It encompasses all aspects of crop production, livestock farming, fishery and forestry. Agriculture is the conversion of solar energy into the chemical energy. Crop production is the conversion of environmental inputs like solar energy, carbon dioxide, water and nutrients in soil to economic products in the form of human or animal food or industrial raw materials.

Agriculture is a science, an art and business of crop production

Agriculture is a science of farming. Scientific principles are employed to find ways of making it as efficient as possible. Through scientific principles plants and animals are transformed genetically and most favorable environment is provided to harvest higher yields of good quality with least expense of energy. The scientific principles of various branches *viz.* soil science, genetics and plant breeding, entomology, plant pathology, microbiology, agricultural engineering *etc.* were employed in agriculture. Agriculture like any other science is a body of truths synthesized and systematized and arranged in such a way as to show the operation of general laws and principles.

Agriculture is an art which embraces knowledge of the way to perform the operation of a farm in a skillful manner. The physical and mental skills are involved in agriculture. The skills may be acquired through years of experience *viz.* ploughing, stacking hay bundles, handling animals, sowing seeds, transplanting, driving a tractor *etc.* Mental skills are those involved in decision making for example when to plough the land, selection of appropriate crops, seed selection *etc.*

Agriculture is a business which is no longer a way of living or subsistence agriculture, where production is intended to meet the home requirements. Agriculture is intended to earn more income. Land, labour, capital are judiciously used. Like in any industry the farming industry should forecast the demand, tailor the production with demand to earn more profits. It involves processing, value addition, transportation, packing and storage in scientific way. Knowledge of employee and employer relationship or human resource management, export and imports, taxation, customs, tariffs and trade are required. All these aspects demand business knowledge in addition to the production and managerial skills. Traditional agriculture is no longer relevant for success in agriculture. Commercial agriculture or corporate farming, agri-business development corporations demand entrepreneurs in agriculture rather than technologists alone.

Importance of Agriculture

Agriculture provides food for human and cattle. raw materials for industries-*e.g.*, textile, jute, sugar, oil, cigarette, tea, coffee, *etc.* Agriculture supports the industries *viz.* fertilizer, pesticides, manufacturing industries–tractor, power tiller, implements *etc.* Agriculture provides employment to large labour class. Some of the agriculture commodities are exported which earn foreign exchange. Traders, transporters and bankers also depend on agriculture.

Agronomy

The word agronomy is derived from the greek words **Agros** meaning 'field' and **nomos** meaning 'manage'.

According to Norman (1980) it is a science of manipulating the crop environment complex with dual aims of improving productivity and gaining a degree of understanding of the process involved.

The recent definition of agronomy is the successful, sustainable, profitable, nutritionally secured, efficient crop production with least or no environmental degradation.

It is a branch of agricultural science which deals with principles and practices of soil, water, and crop management.

Relation of Agronomy with other Sciences

Agronomy deals with methods which provide favorable environment to higher crop productivity.

Agronomy is a synthesis of several disciplines like soil science, Agricultural chemistry, crop physiology, plant ecology, biochemistry and economics. The soil physical, chemical and biological properties have to be understood thoroughly to effect modification of the soil environment.

It is necessary to understand the physiology of crops to meet their requirements. Developments in these subject help in developing new practices which are simpler and economical to provide favorable environment to the crop. The herbicidal properties of 2, 4-D made revolutionary changes in weed control.

Advances in economic analysis helped in production of crops economically. For efficient utilization of resources, systems approach is better than individualistic approach. Instead of taking individual crop into consideration, cropping system as a whole may be considered for taking decisions, *e.g.*, integrated farming system as against cropping alone. Fertilizer recommendation for an entire cropping system saves considerable quantity of fertilizer instead of recommending fertilizer to individual crops.

Scope of Agronomy

India has varied climate and variety of soils. Large human power exists in the country which finds employment in agriculture. Variety of crops can be produced to meet the domestic needs and export.

There exists a large gap between existing and actual yield potential of many crops. Hence there is scope for improving the crop productivity through research and extension activities.

Value addition of the crops is yet another sector to improve farm income. Entrepreneurs have large scope for investment in value addition.

Integrated Farming system, organic farming, adoption of advanced irrigation technology, weed management, production of commercial crops, fodder crops, aerobic paddy production are the potential areas of research, application in the farmers fields or corporates. The knowledge of agronomy is indispensable for augmenting agriculture productivity in the country and the world.

History of Agriculture

Ancient Agriculture

Agriculture has no single or simple date of origin. Man started hunting and gathering food which is most insecure for food. Hunting and gathering food can support one man per square kilometer. He started rearing animals (pastoral stage) which can support three men per square kilometer, then shifted to plant culture which could support 100 men per square kilometer.

The primitive form of agriculture was shifting cultivation, in which man used the crudest tools, cut down a part of the forest, burnt the forest and started growing crops. After a few years when these plots lost fertility, choked with grass or bushes or became heavily infested with soil borne pests, cultivators would shift to new site.

Subsistence agriculture: Grow it and eat it. It is the advanced form primitive agriculture. Hunting and gathering were subsidiary occupations.

Nomadic population followed their flocks and herds in the seasonal pursuit of water, forage and shelter. Commercial ranching blended with subsistence farming. When land became limited shifting cultivation was changed to fallow rotation or legume rotation.

Mixed farming: field crops and animals, field grass husbandry. The same field is used for both cropping and grazing.

During nomadic period man come in contact with useful plants and animals and started domestication.

Right to use land and inheritance of land–land is the security for livelihood. Arable land was divided based on topography. Seasonal cultivation of crops was practiced. During the cropping period the cattle were allowed to graze on a common grazing land in the villages. The byproducts of crops were used as feed for cattle.

Self sufficient agriculture- village was the least common denominator of a permanent society. Later other crafts developed around agriculture. Down trodden people, slaves and bonded labour were the source of human labour.

Man first domesticated sheep and goat (8700-7700BC). Between 7500 to 6500BC man shifted from hunting/gathering food to agriculture. He was using pointed sticks, stone shoes with wooden handles to sow the seeds. Some of the land marks in the history of agriculture are

7500 BC–cultivation of wheat and barley, seeds of wheat and barley were found at Jurmlo in Iraq which dates back to 6750 BC

4400 BC–cultivation maize

3500 BC–cultivation of potato

3000 BC to 1750 BC–Indus valley and Harappan civilization on the banks of river Indus. Huge granaries were excavated in these areas. Evidences of cultivation of barley, wheat, sesame, peas, date palm, cotton, lentil was found. Wooden plough and two wheeled cart were found.

2900 BC–plough was invented

2700 BC–silk moth was domesticated in China.

2300 BC–chickpea, pear, sarson, and cotton cultivation, domestication of buffaloes and foul.

2200 BC–rice cultivation

1800 BC–ragi cultivation

1725 BC–sorghum cultivation

1700 BC–taming of horses

1500 BC–cultivation of sugarcane

Ancient literature *viz.* 'Vedas' have a mention of cultivation of cereals, vegetables, fruits, use of iron implements, ploughing, sowing seeds-broadcasting, fallowing, sequence cropping, use of dung as manure and irrigation.

Civilization flourished on the banks of rivers Nile, Tigris, Euphrates, Indus, and Yellow river (in China). Dams were constructed across the rivers. Aswan dam across the river Nile dates back to 5000 years. Evidences of irrigation dates to 3300 BC.

Greek Agriculture: Hesiod (776 BC) wrote book on crop production.

Herodotus (484 BC to 424 BC) wrote book on date palm cultivation in Babylonia.

Xenophan (430 BC to 330 BC) wrote about ploughing, green manuring (Soil fertility).

Roman agriculture: considered agriculture as a noble profession.
Cato (234 BC to 149 BC) stated the importance of ploughing and green manuring.

Classified the land on utility basis:

1. Land for Vineyard
2. Gardenland
3. Willowland
4. Oliveyard land
5. Meadowland
6. Cornland
7. Small trees and
8. Oak land.

Varro (116 BC to 27 BC): is a soldier cum farmer and scholar. In his book classified three types under agriculture

1. Dealing with production of crops
2. Rearing of large animals like cattle, horses
3. Rearing of small animals like sheep and goat

Virgil (70 BC to 19 BC) stated about crop rotation, inclusion of legumes (soil fertility improvement)

Columella (45AD) wrote book on selection of land, cultivation of crops, plant propagation, and crop improvement.

Romans emphasized on tillage, green manuring, crop rotation.

Early scientific thoughts of agriculture

Peter Discresenzi (1230AD to 1307AD) wrote book "Ruralium Komo Daram". He is referred as father of agronomy.

During the beginning of scientific agriculture, fertilizers were not known. Animal bone powder was used as manure; Guano of South African bird was exported to Britain, Chilian nitrate was exported by South America.

Francis Bacon (1627 AD) stated that the principle requirement of vegetation is water.

Van Helmont (1577–1644 AD) stated that the principle of vegetation is water.

John Woodward (1665–1728 AD) grew spring of mint in distilled and Hydepark conduit shaken with earth. The growth was more in the later. He concluded that the plants absorb terrestrial matter. Earth constitutes vegetation.

Jethro Tull (1674–1740 AD) stated that

- Plants absorb soil particles through their roots.
- He realized the losses caused by the weeds. Invented horse hoe.
- Published new horse hoeing husbandry or an essay on the principles of tillage and vegetation.

Arthur Young (1741–1820 AD) conducted the pot culture experiments to increase the yield of crops by applying several materials like poultry dung, niter, gunpowder *etc*. He published all his findings in 40 volumes as annals of agriculture.

Lord Townshend popular as Turnip Townshend known for Norfolk rotation which comprises clover–wheat–turnip–barley.

1809–Humus theory which is the beginning of soil science.

Sir Humphry Davy (1813 AD) published elements of agriculture chemistry.

Justus Von Liebig (1803–1873 AD) stated that

Mineral theory of plant nutrition–simple analysis of plants and soil will provide reliable index fertilizer needs of the plant.

Law of minimum–the growth of crop is determined by the quantity of what ever the factor which is least in supply in relation to the needs of the plant.

John Bennet Lawes (1814–1900 AD) developed superphosphate by treating rockphosphate with sulphuric acid in 1843. He stated that the phosphorus in bones could be made available by treating with sulphuric acid. J. H. Gilbert (1817–1901 AD) conducted the field experiments in the ancestral field of J. B. Lawes at Rothemstead near Herpenden in Heartfordshire. It is later known as rothemstead agriculture experiment station established in 1843 by Lawes and Gilbert.

The first state funded agricultural experiment station was established near Leipzig in Germany.

Boussingault J.B. (1802–1882 AD) is known as father of field plot experimentation. He established a farm at Alsace in France during 1834 AD for conduct of field experiments. He stated plants obtain mineral nutrients from soil. He also maintained a balance sheet of nutrients taken from soil, rain water and air. He stated that the best crop rotation is one which produces more vegetative matter and the legumes obtain nitrogen from the source not available to other plants.

Laws of heredity and ways to cause mutation by G. J. Mendel (1866 AD) and heredity and inheritance proposed by Charles Darwin (1876 AD) caused major breakthrough in the plant improvement.

Robert Ransome patented a cast iron share in 1785 AD and self sharpening share in 1803 AD.

Efficient seed drill was designed in 1830AD.

Tractor was successfully designed in 1892 AD which made tillage easier. Electricity was used in agriculture during 1920 AD for pumping water from the wells for irrigation.

Large scale manufacture of agriculture implements and machinery took place during 1930AD.

Discovery of Bordeaux mixture by P M A Millerdet (1880 AD) and DDT by Paul Muller (1874 AD) were significant in the field of plant protection.

Development of agriculture in India

During 15th and 16th century AD many crops were introduced to India by Portuguese *e.g.*, Potato, sweet potato, tomato, chillies, pumpkin, papaya, pineapple, guava, custard apple, groundnut, cashewnut, tobacco, American cotton, rubber, sweet and sour oranges, pomegranate, wild brinjal *etc*. During the 20th century groundnut, sunflower and soybean were introduced to India.

Knowledge of agriculture acquired through experience transcended from generation to generation through intimate contacts and co-working with elderly people in the villages. Skills were inherited from generation to generation. Vedas, Vachanas, proverbs, quotes, and poems were the sources of information.

Developments during the 19th century

Botanical gardens were established at Sibpur, Saharanpur, Bangalore and Ootakmand which played role in plant introduction into India from East Indies, China and Europe.

William Carey formed the Royal Agri-Horticultural Societies in Calcutta during 1820, which played role in introduction of crop plants and ornamental plants into India.

In 1891 Dr. J.A. Voelcker submitted report for improvement of indian agriculture and recommended to spread agriculture education in India. Agriculture Chemist to conduct research and Assistant Agriculture Chemist to teach agriculture subjects were posted at Pune, Dehradun and Saidapet.

There were series of famines beginning from 1860 to 1900AD. Among them, the famine of 1896-97 AD is the greatest famine adversely affected the life of 69.5 million people in 3, 07,000 sq miles. Another severest calamity affected 4,75,000 sq miles and 69 million people were affected in almost all the provinces of India. On account of the famines there was acute shortage of food for human, fodder for cattle and raw material for the industry. In 1870 AD a joint department of agriculture, revenue and commerce was established. The famine commission was constituted by the British government comprising of Lord Lawrance Governor and Lord Mayo in the year 1880 AD who suggested the establishment of a separate department of agriculture for increasing the production of food and raw materials for industries.

Developments during 20th century

Agricultural colleges were established at Kanpur, Sabour, Nagpur, Llyalpur (now in Pakistan), Coimbatore during 1906 and at Poona during 1907.

Imperial Agricultural Research Institute was started at PUSA in Bihar during the year 1903 AD. Due to earthquake at PUSA, the institute was shifted to New Delhi in 1936 AD. Later it was renamed Indian Agricultural Research Institute (IARI). Upon the recommendation of Royal Commission on Agriculture, Indian Council of Agriculture Research (ICAR) was established in 1929 AD with its headquarter at New Delhi with the following objectives

- To promote research and education in agriculture, animal husbandry, fisheries and home science.
- To promote extension of knowledge in the field of agriculture, fisheries animal husbandry *etc*.
- To undertake consultancy services in the field of research, education and dissemination of knowledge in the field of agriculture, animal husbandry fisheries to government and other agencies
- To cooperate rural development activities concerning agriculture with other organizations *viz*. Indian Council of Social Science Research (ICSSR), Council of Scientific and Industrial Research (CSIR), Baba Atomic Research Centre (BARC) and other Universities.

ICAR is the apex body responsible for the organization and management of research and education in the fields of agriculture, animal sciences and fisheries in India. The minister of agriculture is the president of the ICAR. Its principal executive officer is director general. He is also secretary to the government of India in the Department of Agricultural Research and Education (DARE)

Research

ICAR has a network of 45 central research institutes, 4 national bureaus 9 project directorates, 30 national research centers, a national academy of agricultural research and management and 77 All India Coordinated Research Projects in the country.

Education

ICAR promotes and coordinates agricultural education in the country through technical and financial assistance to the state agricultural universities postgraduate educational programmes in central institutes and schemes on manpower development and quality improvement.

There are 64 state agricultural universities and three central agricultural universities. These are responsible for promoting agricultural education, research and extension of agricultural technology at the state level. In addition there are four ICAR institutes *viz.* IARI New Delhi, IVRI Izatnagar, NDRI Karnal and central institute for fisheries education Bombay which have been conferred the status of deemed to be universities. These institutes besides carrying out research, impart education at post graduate level in agriculture and animal sciences and also training in the specialized fields.

Some of the major developments prior to and after independence to promote agriculture production in the country are:

1942-Grow more food campaign

1944-Indian central sugarcane committee

1945-Coconut committee

1946-Directorate of plant protection and quarantine committee

1947-Fertilizers and chemicals travencore (FACT)

1947-Oilseeds committee

1954-Punjab Rao Deshmuk founded the Bharath Krishik Samaj- Farmers forum of India

1956-Project for intensification of regional research on cotton, oilseeds and millets (PIRRCOM)

1957-All India Coordinated maize improvement project

Land Grant Universities established with tripartite mission of Teaching, Research and Extension. In 1960 First Agricultural University was established at Pantanagar.

CADA: command area development authority to look into efficient use of irrigation water, soil problems and improving crop productivity in the channel irrigated areas.

1960-Intensive Agriculture District Programme

1963-National Seeds Corporation

1965-Intensive Agricultural Area Programme (IAAP)

1964-National Demonstration Programme(NDP)

1966-High Yielding Varieties Programme(HYVP):

1965-66 Green revolution

1969-Second irrigation committee, Drought Prone Area Programme (DPAP), National Commission on Agriculture (NCA).

1971-All India Coordinated Research Project on Dryland Agriculture (AICRP on DA)

1972-International Crops Research Institute for Semi Arid Tropics (ICRISAT).

1973-Minikit Trials

1973-Department of Agricultural Research and Education(DARE) was formed to link ICAR with centre and State Governments on one hand and international organizations on the other.

1977-Training and Visit System for transfer of agricultural technology to the farmers

1979-National Agricultural Research Project (NARP)

1982-National Bank for Agriculture and Rural Development (NABARD)

1985-National Agricultural Extension project (NAEP)

1986-NARP Phase II

1998-National Agricultural Technology Project (NATP)

2005-National Agricultural Innovation Project (NAIP), Rastriya Krishi Vikasa Yojane (RKVY) was launched

Agricultural revolutions

Green revolution: Dwarf high yielding wheat varieties from Mexico were introduced to India by Dr. N. E. Borlaug (Born 1915), Director of CIMMYT, Mexico. Dr. M. S. Swaminathan (Born 1925) played important role in ushering in the Green Revolution in India. Outstanding contributions were made in basic science of cytology and genetics. There was massive import of Mexican wheat seeds during 1964-67. High yielding varieties coupled with new production technology, use of fertilizers and irrigation revolutionized wheat production which augmented the food production in the country.

Green revolution II: Introduction of Taichung Native-1, first semi-dwarf indica rice variety to India, which revolutionized rice production.

Green revolution III: production of Millets and Maize

White revolution: Milk production, national dairy development board (NDDB) was established in 1970 by Dr. Kurian who was responsible for Operation Flood. Later blue, yellow, round, brown and silver revolution took place which represent the significant increase in the production of fish, oil, potato, biomass and egg production in the country.

Agriculture development in Karnataka

Hebbal Agricultural Farm was established during 1904 in Bengaluru

During 1912-13 Diploma course in Agriculture was offered. Later during1955 LAG (Diploma in Agriculture) was offered.

Agriculture research and development in Karnataka was initiated by Leslie Coleman.

University of Agricultural Sciences Bangalore was established in 1965. At present there are four agricultural universities, one veterinary and animal and fisheries sciences university and one Horticultural Sciences University in Karnataka. Each university has colleges in different parts of the state which impart agricultural education. The zonal agricultural research stations under each university will carryout research to resolve location specific problems. The extension education units (EEU), krishi vignana kendra (KVK), farmers training institutes (FTI), staff training units (STU) were established which will extend the front line technologies developed in various fields of agriculture at the research institutes to the farmers.

Questions

I. Answer the following

1. Justify the statement “Agriculture is an art, science and business of crop production”.
2. What is the importance of agriculture?
3. What is the scope of agronomy?
4. What are the objectives of ICAR?
5. What are the factors of green revolution?
6. State different revolutions of Indian agriculture.

II. Define

1. Agriculture
2. Subsistence agriculture
3. Self sufficient agriculture
4. Shifting cultivation
5. Law of minimum

III. Choose the correct answer

1. Imperial Agricultural Research Institute was established in 1903 at
   (a) PUSA in Bihar (b) Pantanagar
   (c) New Delhi (d) Hyderabad
2. Horse hoeing husbandry was published by
   (a) Liebig (b) Jethro Tull
   (c) Boussingault (d) Von Helmont
3. Mineral theory of plant nutrition was stated by
   (a) Lawes (b) Gilbert
   (c) Liebig (d) Darwin
4. An international research intstitute
   (a) IIHR (b) CRRI
   (c) IARI (d) ICRISAT
5. Agriculture is
   (a) An art and science of crop production
   (b) A business of crop production
   (c) Science and business of crop production
   (d) An art, science and business of crop production
6. ICAR was established in the year
   (a) 1965 (b) 1929 (c) 1903 (d) 1935
7. Yellow revolution refers to
   (a) Egg production (b) Biomass production
   (c) Oil seed production (d) Fish production
8. Rothemsted agriculture research station was established by
   (a) Jethro Tul (b) Justus Von Liebig
   (b) Lawes and Gilbert (d) J. B. Bousingault
9. The person responsible for operation flood in India was
   (a) Dr. M. S. Swaminathan (b) Dr. N. E. Borlaug
   (c) Smt. Indira Gandhi (d) Dr. Kurian
10. The word "Agriculture" is derived from
    (a) Greek (b) Roman (c) Sanskrit (d) Latin

11. Father of Agronomy

    (a) Jethro Tull (b) Peter Decrescenzi
    (c) Blackman (d) Liebig

12. Greek historian who mentioned about ploughing

    (a) Xenophan (b) Virgil
    (c) Van Helmont (d) Arthur Young

13. Cultivation of sugarcane dates back

    (a) 2200 BC (b) 1500 BC
    (c) 250 BC (d) 2900 BC

Chapter 2

Factors Affecting Crop Production

Crop growth is influenced by internal factors and external factors. Internal factors are controlled by the genes and hereditary. External factors are climate, edaphic, biotic, physiographic and anthropic.

Genetic Factors

Genetic factors are controlled by the genes. Plant height, yield, size and color of the grains or fruit *etc*. The productivity and quality of the crop can be improved by breeding through conventional methods or by biotechnological approaches.

Environmental factors

Climate

Precipitation occurs in the form of rainfall, snow, hail and fog.

Fog: Consists of water droplets so small that their fall velocity is negligible. Fog particles contact vegetation and may adhere, coalesce with other droplets and eventually form a drop which is large enough to fall to ground.

Dew: During night there is loss of heat by radiation. Condensation of water vapour present in the air results in dew. Winter crops are efficient collectors and users of dew.

Rainfall is the most important factor affecting the vegetation of place. Most of the crops receive their water supply from rain. The yearly precipitation *i.e.*, quantity, intensity and distribution largely influence crop growth.

Low and ill distributed rainfall is most common in dry lands wherein drought tolerant crops are grown.

Heavy and regular rainfall is the common feature in western ghats. High water requiring crops are grown *e.g.*, coffee, cardamom, pepper, banana *etc.*

Desert: it is the least rainfall receiving area. Desert grasses and shrubs are common vegetation.

Drought: is condition of continuous lack of moisture so serious that crops fail to develop and mature properly

Adaptation to moisture situations

Plants assimilate 0.1-0.3 percent of the water absorbed from the soil.

Hydrophytes: Aquatic plants which are grown in water. Swamp and bog plants. They have spongy tissue. Stomata are numerous and located on the upper side of the leaves *e.g.*, Waterhyacinth, rice, eelgrass.

Mesophytes: Most common land plants. Stomata are more confined to or more numerous on the under side of the leaves. Root hairs are abundant. Root length and volume often equal or exceed top growth. True mesophytes wilt after losing 25 percent of their total water content. Xerophytic- mesophytes wilt after losing 25 -50 percent of their total water content.

Xerophytes: These are capable of enduring prolonged drought without injury. They will grow in a substrate which is depleted of water for growth to a depth of 20 to 25cm. They have modifications *viz.* reduced stomata, respiration, transpiration. In cati the carbon dioxide released during respiration stays in chlorenchyma tissue. It is assimilated. Hydrophilic colloids in certain wheat varieties induce drought resistance due to presence of bound water. During drought they expend not more than 2–10 percent of the water absorbed.

Effects of excessive moisture

- Limit oxygen supply
- Formation of toxic substances
- Leaching of nutrients *e.g.*, nitrates
- Detrimental to germination, flowering, pollination and fruiting
- Continued turgidity, low transpiration,
- Disease incidence *viz.* rust, mildews,
- Curing and storing of produce hindered

Temperature

Temperature influences the following plant processes

- Biochemical reactions
- Uptake of carbon dioxide
- Production of chloroplasts
- Production of growth substances
- Photosynthesis
- Dry matter production
- Germination
- Leaf initiation
- Leaf emergence

- Leaf expansion
- Flowering
- Spikelet development
- Grain development
- Yield

Various biochemical process are associated with photosynthesis are controlled by temperature. Under high temperature photosynthesis become heat inactivated. Retardation of growth and adverse effects on fertilization at temperature below lethal limit are observed.

Temperature is the measure of heat energy. The range of maximum growth for most agricultural plants is between 15-40°C. Temperature of a place is largely determined by the distance from the equator (latitude) and altitude. Vegetation is classified into four classes based on the temperature requirement.

Megatherms: Equatorial and tropical, high temperature requiring *e.g.*, rice, rubber, banana

Mesotherms: Tropical and subtropical, high temperature alternated with low temperature, *e.g.*, maize, sorghum

Microtherms: Temperate and high altitude plants, low temperature requiring, wheat, oats, potato

Hekistotherms: Very low temperature requiring *e.g.*, Pines, spruce
Every plant community has its own minimum, optimum and maximum temperature known as cardinal points.

Cordinal temperature of certain plants (°C)

Crops	Minimum	Optimum	Maximum
Wheat	3.8–4.4	24.9	29.9–32.2
Barley	3.8–4.4	19.9	27.7–29.9
Oats	3.8–5.0	24.9	29.9
Maize	7.7–9.9	31.6	39.9–43.8
Sorghum	7.7–9.9	31.6	39.9
Rice	9.9–11.6	32.2	36.1–38.3
Tobacco	12.7–13.8	27.7	34.9

The minimum daily mean temperature at planting time for potato 7°C, corn 14°C, Cotton 17°C. minimum temperature for germination of maize 4°C, sorghum 9°C, rice 9°C,

Minimum temperature for growth initiation for sorghum 15–18°C, optimum temperature for most temperate crops is 24–29°C, maximum temperature 35–41°C

Cool season crops fail to grow at an average temperature of 30–38°C

Wheat, potato, barley, oats require max temperature of 30–38°C, min 0–5°C, optimum temperature 20–30°C

Warm season crops *viz.* sorghum, maize, sugarcane, groundnut, redgram cowpea, pearlmillet require maximum temperature of 40–50°C, minimum 15–20°C, optimum temperature 30–38°C.

Temperature effects on plants

Chilling injury: some plants growing in hot climate if exposed to low temperature (above freezing point) express chloratic condition or bands on the leaves. *e.g.*, sorghum, sugarcane, maize when exposed for 60 hrs at 2–4°C, on the other hand cold loving plants *viz.*, potato, sunflower and tomato are unaffected.

Freezing injury: this is generally caused in plants growing in temperate regions. Water is frozen into ice crystals in the intercellular spaces. Frost damage in potato and tea is common.

Suffocation: during winter ice or snow form a thick cover over the ground and the crop suffers for want of oxygen.

Heaving: injury to plants caused by a lifting upward of the plant along with the soil from its normal position in temperate regions where snow fall is common.

Heat injury: very high temperature often stops growth. The plant faces incipient starvation due to high respiration rates. The plant is stunted. If such condition persists for a long period the plant is killed. Sterility of plants, young seedlings are killed, defoliation premature dropping of fruits are the adverse effects of high temperature.

Vernalization: some plants require cold stimulus before they come for flowering. The cold treatment given to the sprouting seeds to effect the flowering is known as vernalization. Lysenko a Russian scientist proposed the incubation of sprouted seeds at temperature just above 0°C for 2–3 weeks before sowing.

Thermoperiodism

The response of plants to rhythmic fluctuations in temperature is known as thermoperiodism. A number of physiological process *viz.* germination, stem elongation, fruiting, floral development and increase in frost hardiness may proceed at most satisfactory rate under rhythem of alternating temperature.

Humidity: refers the invisible water vapour present in the atmosphere. The air is said to be saturated when it holds maximum amount of water vapour at a particular temperature. The humidity in atmosphere is termed as relative humidity (RH)

Importance of relative humidity

It is related to water relations in plant. Directly related to evapotranspiration.

Indirectly related to leaf growth

Photosynthesis: when RH is low transpiration increases causing water deficits in the plant. Water deficit causes partial closer of the stomata and increases the mesophyll resistance thereby blocking the entry of carbon dioxide.

Pollination: when RH is high pollen dispersal from the anthers is affected. Seed set is more at moderate RH than at high RH.

Pests and disease: Incidence of pests and diseases is high under high humidity. High relative humidity favours germination of fungal spores. Spread of blight disease of potato and tea is more rapid under high RH. Aphid and jassid incidence is more under high RH.

Wind: the turbulence in the atmosphere is termed as wind. Moderate winds are essential for pollination, exchange of carbon dioxide in the canopy, winds also cause rains.

Adverse effects of wind

High wind velocity increases transpiration, accelerates the desiccation of crop reduce plant height, normal form and position of the shoot is permanently deformed when developing shoot is continuously exposed to wind from particular direction. Lodging of field crop, tearing of leaves, dropping of fruits, grain shedding, soil erosion, root exposure in deserts, spread of insect pests, spores of fungi. Wind also alters the balance of hormones in plants- wind increases ethylene production in barley and rice. Wind increases gibberilic acid content of roots and shoots in rice. Nitrogen concentration in both barley and rice increases with increase in wind speed.

Atmospheric gases

Nitrogen of the atmosphere is directly used by symbiotic and asymbiotic nitrogen fixers. Carbon dioxide used by the plants during photosynthesis. Oxygen required for respiration. The concentration of nitrogen, carbon dioxide and oxygen in air is 78.19, 0.03 and 20.95 percent respectively. Sulphur dioxide and nitric oxide in atmosphere reach soil during rains improve soil fertility. Carbon dioxide enrichment of canopy improves crop productivity. On the other hand increase in concentration of carbon dioxide, methane, nitrous oxide and fluorochlorocarbons in the atmosphere deplete ozone layer causing ultraviolet rays to reach earth and causes global warming.

Light

Light is one of the most important factors influencing many vital plant processes. It is required for synthesis of chlorophyll pigment which is capable of absorbing radiant energy and converting it into potential chemical energy *viz.* carbohydrates through the process called photosynthesis. The photosynthesis is directly proportional to the amount of light. Other processes which are influenced by light are seed germination, leaf expansion, growth of stem and shoot, production of tillers, branches, flowering, fruiting, root development growth movements in plants.

Under low light intensity plants grow tall with weak stem which may cause lodging.

Plants obtain light from solar radiation which is the source for light and heat. Light efficiency in plants is less than two percent.

The quality (wavelength and colour) the quantity (the intensity and duration of exposure to light) greatly influence the plant growth.

Quality refers to the wave length and colour. Of the total range of electromagnetic wavelengths in the solar spectrum, light or the luminous energy includes wavelengths between 400-750mμ (milli microns) or nm (nanometers). Lux is the unit of illumination. It is the light intensity from a standard candle at one metre distance is a metre candle or lux.

Quality of light and its effect on plant growth

Colour	Wave length	Effect on plants
Ultraviolet	< 390 mμ	X rays and gamma rays. Very detrimental for growth accounts 0-4%
Violet	400–435 mμ	Favorable for plant growth, Phototropism
Blue	435–490 mμ	Favorable for plant growth. Photosynthesis, blue is more efficient than red.
Green	490–574 mμ	Carbon dioxide assimilation
Yellow	574–595 mμ	Carbon dioxide assimilation
Orange	595–626 mμ	Carbon dioxide assimilation
Red	626–750 mμ	Carbon dioxide assimilation,more favorable for plant growth
Infrared	> 750 mμ	Temperature will increase

Photoperiodism

Flowering requires a definite period of vegetative growth. It varies from plant to plant. Physiological mechanism responsible for flowering has been found to be controlled by periodicity of light (photoperiodism). Growth and development of the plant is controlled by intensity and quality of light. Length of the day light period has been found to have marked influence on the behavior of the plant particularly flowering. Garner and Allard (1920) observed with tobacco variety Marryland Mammoth flowered when the relative length of the day was shorter than the length of the dark period lead to the discovery of photoperiodism. Photoperiod is the response of the plants to the relative length of the day and night and photoperiod to designate the favorable length of the day for each plant. On the basis of length of photoperiod requirements plants have been classified as short day plants, long day plants and day neutral plants.

Short day plants: For flowering of short day plants the day length must not exceed a certain critical length. They are more critically called long night plants. A certain uninterrupted dark period of 24 hours is necessary for their flowering. If the dark period is less than certain critical length flowering will not occur. Flowering can be induced if the plants even in long days by providing the dark period *e.g.*, soybean, potato, sugarcane, tobacco, redgram and niger.

Long day plants: long day plants require a photoperiod of more than a critical length which may vary from 14-18 hours. The best flowering of long day plants usually occurs in continuous light. For flowering they require no dark period or a very short dark period *e.g.*, Radish, alfalfa, sugarbeet, opium popy, oats, maize, wheat.

Day -Neutral plants: flowering is not affected by the length of the day *e.g.*, Sunflower, cotton, pea, tomato, cucumbers.

Significance of photoperiodism: short day and long day plants should be planted or sown to coincide the flowering period to either short day or long day respectively. Hence they are season bound. They cannot be grown throughout the year. Day neutral plants can be grown in all the seasons. They are not season bound.

Physiographic factors

The nature of the surface of earth is known as topographic factors. They include

- Altitude of a place
- Steepness of slope
- Exposure of the slope to light and wind
- Direction of mountain chains

The topographic factors affect the crops indirectly by modifying the climate and edaphic factors of a place.

Altitude of a place: Increase in altitude is usually associated with

- Decrease in temperature
- Increase in precipitation
- Increase in wind velocity

Temperature decreases at the rate of 6-7°C per 1000 meters except in the valleys and low lands due to the downward descent of cold air. Due to increased rate of precipitation and strong winds, the type and nature of soil and vegetation are affected. The organic matter content of the soil increases and this in turn increase the N content and acidity of the soil.

Steepness of slope

- Decreases the moisture content of the soil
- Causes run off and soil erosion

Exposure of slope to light and wind

- Northern slopes of temperate regions and Himalayas–exposed to weak intensity of light and strong dry winds have poor crop
- Western slopes in hilly areas of Tamil Nadu have poor crops due to damage by high velocity winds
- Southern slopes of the temperate region of Northern Hemisphere will receive high intensity of light. Incident rays will be perpendicular. So crop growth will be good.

Direction of mountain chains governs rainfall of a place which in turn determines the type of crop.

Edaphic Factors

Soil

Functions of soil

- Soil provides mechanical support to the plant: Roots are anchored to the soil enable growing plant to remain erect.
- Soil supplies mineral nutrients
- Soil colloids hold the nutrients due to presence of charge on their surface
- Soil holds water. Water held in the soil is called soil moisture. It is held with cohesive and adhesive forces operating in the capillary pores
- Soil supplies oxygen required for respiration of roots and soil organisms
- Retains heat and maintains the temperature required for the growth of plant root and soil organisms

The depth of the soil, soil texture, soil structure, extent of soluble salts or acidity, water holding capacity, cation exchange capacity, influence the crop growth.

Functions of water

It is the medium for plant growth and source of hydrogen and oxygen.

Water is required for various physiological functions. Constituent of protoplasm, maintains the turgidity of cells, responsible for cell elongation, water is a solvent for mineral nutrients. Water is required for movement of the guard cells, folding of leaflets, opening of flowers.

Agronomically water is required for seed bed preparation, sowing the seeds, seed germination, emergence of seedlings. Water has special roles in protecting the crop against frost damage, high temperature, harvesting the crops *viz*. groundnut, potato, and for land reclamation.

Gaseous phase: Soil air and water are interrelated. Soil air is rich in carbon dioxide (>0.03%) and low in oxygen (5-10% as against 20 percent in atmosphere).Too high concentration of Carbon dioxide in the soil cause toxic effects on plant roots. There should be free exchange of gases between soil and atmosphere to ensure free supply of oxygen to plant roots.

Soil temperature: The temperature in the 0-30cm soil depth from the surface influences crop growth. Optimum soil temperature promotes root growth, absorption of nutrients, activity of microorganisms, and nutrient transformations. Soil temperature of 29-33°C is optimum for most of the plants. Very high soil temperature 48-53°C is harmful for germination of seeds, seedling growth, there will be loss of nutrients by volatilization and accelerated rate of organic matter decomposition, loss of soil moisture. Very low soil temperature (<20°C) on the other hand reduces the activity of roots and root growth, germination of seeds, activity of microorganisms. Organic matter turnover is reduced, nutrient mobility and absorption by the plant roots will be reduced.

Soil reaction or soil pH is the symptom of soil condition or soil health. The cause is Al^{3+}, H^+ or OH^-ions. Soil pH normally varies from 4.5 to 8.5

Soil pH category	pH
Normal soils	6.5–7.2
Strongly acidic soils	< 5.0
Moderately acidic soils	5.0-6.5
Saline soils	< 8.5
Sodic soils/Alkaline soils	> 8.5

Soil reaction greatly affects the availability of most of the plant nutrients. Soil reaction or pH greatly affects the availability of most of the plant nutrients. At pH < 5 Fe, Al and Mn become more soluble and toxic. As the pH increases towards neutrality (moderate acidity) their solubility or availability is not so deprived. They will not develop toxic limit. If pH is more than 7 plants develop deficiency symptoms. Cu, Zn and Co also available at low pH values. At pH 7 they become critical. The availability of nitrogen, sulphur and molybdenum are somewhat restricted at low pH values, the availability of phosphorus is the best at intermediate pH values. Ca and Mg availability increases with increase in pH except in soils which have appreciable amount of Na. Even in degraded alkali soils increase in pH shows presence of Ca and Mg. Boron availability is more in the pH range 5 to 7. At pH < 5, Boron availability decreases between pH 7 and 8.5. Again when pH >8.5 B availability increases. Between 7 and 8.5 there is fixation of B by Ca which hinder the movement and metabolism of B. Molybdenum is not available in strongly acid soils (at pH 4). As the pH increases from 6 and above availability increases. Chlorine is unavailable in alkaline soils. The favourable pH for most of the crops is between 6 and 7.

Soil reaction also influences the activity of microorganisms. The general purpose bacteria and actinomycetes function best at the intermediate and high pH values. Fungi seem to be versatile, flourish satisfactorily over a wide pH range. Actinomycetes predominate in acid soils, whereas at intermediate and at high pH there may be stiff competition from bacteria and actinomycetes.

Biotic factors

Competitive and complementary nature among field crops: competition between plants occurs when there is demand for nutrients, moisture and sunlight. Closer spacing reduces yield of individual plants, wider spacing will reduce the total yield because of lesser population of plants per unit area. Hence optimum spacing is an important agronomical practice. But in case of different crops when grown together as in the case of mixed cropping, there is mutual benefit resulting in better yield eg cereal- legume mixture.

Competition between weed and crop: weeds are plants growing where they are not wanted. Weeds reduce crop yields on account of their competition with crops for water, soil nutrients and light. The increase in cost of labour and equipment, reduce the quality and marketability of agricultural produce, harbour insect pests, fungal pathogens, virus and bacterial organisms and some are poisonous to human beings and livestock. In non-irrigated areas the competition is mainly for water and in irrigated areas the competition is for nutrients. Weeds in irrigation channels and drainages impede the flow of water.

Plants as parasites: a plant parasite is dependent on its host plant for its existence *e.g.*, bacteria, fungi and viruses.

Saprophytes: these microorganisms obtain their food either by breaking down dead plant and animal remains.

Parasites: they attack living plants and animals. They cause disease

Symbiosis: the biological interrelationship among the organisms is termed as symbiosis.

e.g., symbiotic relationship between legumes and rhizobia results in N fixation.

Free living bacteria (Azotobacter)

Animals

Animals living in soil: the soil fauna including protozoa, nematodes, rotifiers, snails and insects. They help in using organic matter. Insect pests cause damage to the crops. The average loss caused by insects amounts to 20 percent.

Beneficial organisms: pollinators, earthworms

Small animals: rats, rabbits, squirrels *etc.* damage the crop

Large animals: cattle, goat, sheep, donkeys, horses, deer, elephants, wild pigs often damage the crops

Anthropic factors

Man has produced profound changes in plant environment both by genetic improvement and crop management practices.

- Plant improvement by plant breeders
- Introduction of exotic plants
- Scientific soil and crop management practices
- Adaptation of economically profitable husbandry methods

Questions

I. Answer the following

1. What are the forms of precipitation?
2. What are the types of plant adaptation to soil moisture?
3. What are the ill effects of excessive soil moisture on crop plants?
4. What is the significance of temperature in plant growth?
5. Classify the plants based on the temperature.
6. What are the physiographic factors influencing crop growth?
7. What are the functions of soil in relation to the plant growth?
8. What are the biotic factors influencing crop growth ?
9. How does anthropic factors influence crop distribution?

II. Define

1. Cordinal temperature
2. Vernalization
3. Thermoperiodism
4. Short day plant
5. Long day plant

III. Underline the correct answer

1. An example of hydrophyte
   (a) Agave (b) Rice
   (c) Sugarcane (d) Maize
2. An example of day neutral plant
   (a) Wheat (b) Maize
   (c) Bengalgram (d) Redgram
3. Vernalization refers to
   (a) Sprouting of seeds
   (b) Cold treatment given to the sprouting seeds
   (c) Dring of seeds in hot sun
   (d) Seed hardening
4. An ideal pH for plant growth
   (a) 8.0 (b) < 4.0
   (c) 6.5–7.2 (d) 4.5–5.5
5. Optimum temperature for plant growth is
   (a) 24–29°C (b) 15–20°C
   (c) 35–41°C (d) 8–10°C
6. Banana is a
   (a) mesotherm (b) megatherm
   (c) microtherm (d) hekistotherm
7. One of the following crop is not adversely affected when exposed to the temperature 2–4°C
   (a) Sorghum (b) maize
   (c) sugarcane (d) potato
8. Niger is
   (a) long day plant (b) phototropic
   (c) short day plant (d) day neutral plant
9. Optimum soil temperature for plant growth
   (a) 29–33°C (b) 15–20°C
   (c) 48–53°C (d) 40–48°C

10. Optimum soil pH for plant growth
    (a) 6.5–7.2 (b) 4.5–5.5
    (c) 7.0–8.5 (d) 5.5–6.5
11. Yield of individual plant is high under
    (a) optimum plant population (b) High plant population
    (c) low plant population (d) Very high plant population
12. Bacteria and actenomycetes population is high under
    (a) neutral to basic soil pH (b) acidic soil pH
    (c) strongly acidic pH (d) at pH above 8.5
13. Banana plant is a
    (a) mesotherm (b) megatherm
    (c) microtherm (d) hekistotherm

Chapter 3

Classification of Field Crops

Field Crops: Refers to the crops grown on a vast scale. They are classified in the following ways

1. According to the place of origin
2. Botanical classification
3. Commercial classification
4. Economic classification
5. Seasonal classification
6. Classification according to ontogeny
7. Agronomic classification
8. Classification based on leaf morphology
9. Classification based on serving special purpose

1. According to place of origin

(a) Native crops- *e.g.*, rice, wheat, sugarcane

(b) Exotic or introduced crops–groundnut, sunflower, potato, tobacco, maize

2. Botanical or Taxonomic classification

Crops are grouped under different families based on morphological, anatomical and physiological similarities

(a) Poaceae (Graminae): Rice, Wheat, Oats, Jowar and Sugarcane

(b) Papilionaceae (Syn: fabaceae): All legumes, Groundnut and Redgram

(c) Cruciferaceae: Mustard and Toria

(d) Malvaceae: Cotton and Bhendi

(e) Solanaceae: Potato, Tomato, Brinjal and Tobacco

(f) Tiliaceae: Jute

(g) Asteraceae (Compositae) : Sunflower, Safflower and Niger

(h) Chenopodiaceae: Sugarbeet and beetroot

(i) Pedaliaceae: Sesame

(j) Enphorbiaceae: Castor

(k) Convolvulaceae: Sweet Potato

(l) Umbelliferae: Coriander, Cumin, Carrot

(m) Liliaceae: Onion and Garlic
(n) Zingiberaceae: Ginger and Turmeric

3. Commercial classification

(a) Food crops: Cereals and pulses
(b) Feed crops: Grasses and fodder crops
(c) Commercial crops: Cotton, Sugarcane, Tobacco and Jute
(d) Food adjuncts: Turmeric, Cumin and Garlic

4. Economic or agrarian or agricultural classification or based on the use of crop plants and their products.

(a) **Cereals:** 'Ceres' (A Roman word) derived from the name of a Roman goddess who was the 'giver of grain'. They contain carbohydrates which are staple food.

e.g: Rice, wheat, maize, sorghum

(b) **Millets:** Small grains, They have a single cover (due to fusion of testa to the pericarp). Based on the size of the grain they are classified as major millets and small millets.

Large millets/major millets: Sorghum, bajra

Small millets/minor millets/nutri millets: Finger millet, small millet, proso millet, kodo millet and barnayard millet

(c) **Oil seeds**: Rich in fatty acids which are used to extract vegetable oil *i.e.*, edible and nonedible oil.

(d) **Pulses:** On splitting they produce dalwhich is used as food.

(e) **Feed or forage**: Hay, silage, soilage, pasture fodder.

e.g: Anjan grass, Berseem, Centro grass, Guinea grass, Lucerne, Rhodes and Styloxanthus sps.

(f) **Fibre crops**: Seed fibre: Cotton

Stem fibre or bark fibre: Jute, Roselle fibre

Leaffibre : Agave

(g) **Sugar and starch crops:** Sugarcane, Sugar beet, Potato and Sweet Potato

(h) **Spices and condiments:** They are used to season the food, impart flavour, taste, add zest and colour.

i. **Spices:** which give aroma and flavour.

ii. **Condiments:** Impart a relishing taste.

e.g: Ginger, Garlic, Fenugreek, Cumin, Turmeric, Chillies, Onion, Coriander, Anise and Asafoetida

(i) **Drug:** Tobacco, Mint

(j) **Narcotics, fumitories and masticatories**

They have stimulating, numbing (Deprive sensation) drowsing or relishing effects upon consumption. *e.g.*, Tobacco, Ganja, Opium poppy and anise.

(k) **Beverage:** Tea, Coffee Cocoa

(l) **Rubber:** Hevea brasiliensis

(m) **Green leaf manure:** glyricidia, pongamia

(n) **Green manure :** Crotalaria juncea, Dhaincha

5. Based on ontogeny: According to life cycle

(a) **Annual crops:** Completes the life cycle within a season or year

(b) **Biennial crops:** Crop plants having a life span of two consecutive seasons or years.

e.g: sugarbeet, beet, radish and carrot.

(c) **Perennial crops**: Sugarcane, napier, ginger and sweet potato

6. Seasonal classification

(a) **Kharif or south west monsoon crops:** Crop plants grown during June - July to September–October such as rice, maize castor, groundnut. They require warm wet weather during their major growth period and shorter day length for flowering

(b) **Rabi crops:** Crop plants grown during October–November to January–February such as wheat, mustard, barley, oats, bengalgram (chickpea), peas, linseed, lentil, cabbage and cauliflower. They require cold–dry weather for their major growth period and longer day length for flowering.

(c) **Zaid or summer crops:** Crops plants grown during February-March to May-June such as black gram, green gram, sesame and cowpea. They require warm dry weather for their major growth period and longer day length for flowering.

In rice seasonal classification is refered as aus, aman and *boro*.

(a) **Aus:** Derived from Sanskrit word *ashu* meaning quick or early. It is designated to the rice varieties grown during pre-monsoon period. The rice is known as autumn rice. It is harvested in Aug-Sept.

(b) **Aman:** It is derived from Arabic meaning 'safety'. Synonymous to Kharif or winter crop of rice.

(c) **Boro:** Refers to the rice crop grown in submerged land lower in elevation. Grown during January-February to April-May. It is known as summer rice.

Drawbacks of seasonal classification:

- It is not universal

- Some crops grown throughout the year. *e.g.*, photoinsensitive crops like sunflower
- Some crops overlap the seasons *e.g.*, Jute
- Some crops do not follow season *e.g.*, Cotton, pigeonpea.

7. Classification according to cultural requirements or agronomic classification

A group of crop plants are alike in cultural requirements due to their similar agro-botanical or morpho-agronomical character.

(a) **According to suitability of toposequence**

i. Upland crops: pigeonpea, groundnut, sunnhemp, maize, jowar and bajra

ii. Medium land: Potato, sugarcane, ragi, wheat, bengalgram, barley, oats, mustard and upland rice.

iii. Low land crops: Rice, dhaincha and paragrass

(b) **According to the suitability of textural groups of soils**

i. Sandy to sandy loam (light) soils: Potato, sweet potato, sugarbeet, onion, carrot, turnip and greengram.

ii. Crops of silty or silty loam (medium) soils: Jute, sugarcane, maize, mustard and tobacco.

iii. Crop of clay to clay loam (heavy) soils: Rice, wheat, barley, linseed, lentil, paragrass, guineagrass and marvel grass.

(c) **According to the tolerance to the problematic soils**

i. Crops tolerant to acidic soils: submerged paddy, potato and mustard.

ii. Crops tolerant to saline soils: Wheat, chillies, cucurbits, bajra, jowar and barley.

iii. Crops tolerant to alkaline soils: Barley, cotton and chickpea.

iv. Crops tolerant to water logged soils: submerged paddy, dhaincha, paragrass, napier and guineagrass.

v. Crop tolerant soil erosion: marvel grass, dhaulu, groundnut, blackgram and kudzu.

(d) **According to tillage requirement**

i. Arable crops: Need preparatory tillage. *e.g.*, sugarcane, potato, maize.

ii. Non-arable crops: Do not need preparatory tillage. *e.g.*, blackgram, cowpea and paragrass.

(e) **According to depth of root system**

i. Shallow rooted: onion, rice and potato.

ii. Moderately deep rooted: wheat and groundnut.

iii. Deep rooted crops: cotton and sweet potato.

iv. Very deep rooted crops: sugarcane and pigeonpea.

(f) **According to tolerance to hazardous weather conditions**

i. Frost tolerant crops: Sugar beet and beet.

ii. Cold tolerant crops: Potato, cabbage and mustard.

iii. Drought tolerant: Bajra, jowar, barley, safflower, castor and cotton.

(g) **According to water supply**

i. Irrigated crops: potato and cabbage.

ii. Rainfed crops: upland rice, jowar and bajra.

iii. Residual but conserved moisture crops: chickpea, greengram and blackgram.

(h) **According to method of sowing or planting**

i. Direct seeded crops: wheat, maize, groundnut and legumes(redgram, blackgram).

ii. Planted: Sugarcane, sweet potato and potato.

iii. Transplanted crops: Rice, onion and ragi.

(i) **According to requirements of inter tillage specially earthing up**

i. Inter tilled: maize, potato and groundnut.

ii. Non inter tilled: fodder jowar and paragrass.

(j) **According to duration of the crops**

i. Very short duration (75 days): green gram, blackgram and radish.

ii. Short duration (75 to 100): upland rice and potato.

iii. Medium duration (100 to 125 days): rapeseed, wheat, jowar, bajra and Barley.

iv. Long duration (125 to150 days): cotton and rice.

v. Very long duration (150 days and above): redgram and sugarcane.

8. Classification based on leaf morphology

i. Broad leaved: mustard, cotton and tobacco.

ii. Narrow leaved: rice and ragi.

9. Special purpose crops: Grown with the intention of serving purposes in addition to obtaining yields or to meet a particular purpose.

(a) **Catch crops or emergency crops or contingent crops:** These are crops cultivated to 'catch' the forthcoming season. They replace the main crop that has failed due to biotic or climatic or management hazards. These are short duration crops with quick growing ability.

e.g: green gram, black gram, cowpea, bajra and coriander.

(b) **Restorative crops**: They restore or ameliorate the soil. *e.g.*, legumes.

(c) **Exhaustive crops**: They exhaust the soil nutrient. *e.g.*, sunflower, maize linseed, sesamum and brinjal.

(d) **Paira crops:** Sown before harvesting the main crop.

(e) **Smother crops:** They suppress the weeds by their dense foliage due to their quick growing ability, tillering or branching habit. *e.g.*, cowpea and mustard.

(f) **Cover crops:** These crops protect the soil surface from erosion and they have thick foliage or root mats *e.g.*, groundnut, para grass and rice bean.

(g) **Nurse crops:** They nurse other crops by providing shade or acting as climbing sticks.

e.g., *Erythrina*–shade coffee.

Agase–Subabul–provide support to betelvine.

Arecanut plant- provides physical support to peppervine.

Sesbanea- provides support for betelvine.

(h) **Guard or barrier crops:** These crop plants help to protect another crop plants from trespassing or restrict the wind speed.

e.g., Safflower sown to protect gram.

Mesta sown along the border of sugarcane.

(i) **Trap crops:** Grown to trap soil borne harmful biotic agents such as parasitic weeds. They stimulate germination but fail to support further growth.

e.g., cowpea stimulates striga seed germination. Striga is parasitic weed in sorghum.

Trap insects: Mustard in cabbage field traps cabbage borer.

(j) **Mulch crops:** They provide a thick covering on the ground and helps to conserve soil moisture. *e.g.*, cowpea.

(k) **Sod or Turf crops:** Belongs to grass family. They have sod type tillers with matted foliage and roots close to the soil surface. They are grown to conserve soil from erosion particularly in non-arable areas *e.g.*, marvel grass, *Cynodon dactylon* and *Digitaria sanguinalis*.

(l) **Cleaning crops:** They clean the field from weeds *e.g.*, potato and groundnut.

(m) **Fouling crops:** They allow infestation of field by weeds *e.g.*, maize, onion, cotton and upland rice.

(n) **Cash crops:** For sale and earn cash. They need processing.

(o) **Ware crops:** These crops are temporarily stored in ware houses for future use or sale. They include fleshy fruits or roots, tubers of high value crops. *e.g.*, Potato, Beet root and Carrot.

(p) **Truck crops:** Marketed in fresh condition. *e.g.*, ladies finger, brinjal and spinach.

(q) **Cole crops:** Cold weather requiring crops *e.g.*, cabbage and cauliflower.

(r) **Silage crops:** They are preserved in pits (*Silo* pits) in a succulent condition by a process of natural fermentation or acidification for feeding livestock during lean months or off seasons. *e.g.*, maize, cowpea and fodder jowar.

(s) **Soiling crops:** These are crops harvested while they are still green and fed fresh to livestock in stalls *e.g.*, Berseem, teosinte, napier and oats.

(t) **Riparian crops:** Grown along irrigation or drainage channels or water bodies to protect soil from erosion *e.g.*, field bind weed, paragrass and babul.

Questions

I. Answer the following

1. Classify the field crops according to agriculture or based on the use of their products.
2. What is seasonal classification of crops?
3. Classify the crops based on the agronomic requirements.
4. What are the special purpose crops? List them with one example for each.

II. Define

(a) Cereal
(b) Millet
(c) Narcotic
(d) Aus crop
(e) Catch crop
(f) Nurse crop
(g) Trap crop
(h) Truck crop
(i) Soiling crop

III. Underline the correct answer

1. An example of a rabi crop

   (a) rice (b) bengalgram
   (c) groundnut (d) ragi

2. An exhaustive crop
   (a) maize (b) soyabean
   (c) groundnut (d) sunnhemp
3. An example of deep rooted crop
   (a) castor (b) onion
   (c) potato (d) coriander
4. Maize is a
   (a) cereal (b) pulse
   (c) oilseed (d) pseudocereal
5. Crop of low lands
   (a) paddy (b) sunflower
   (c) soyabean (d) green gram
6. Jute is classified as
   (a) beverage crop (b) pulse crop
   (c) spice crop (d) fibre crop
7. One of the following is not an indigenous crop
   (a) rice (b) sunflower
   (c) wheat (d) sugarcane
8. Onion belongs to
   (a) solanaceae (b) umbelliferae
   (c) liliaceae (d) malvaceae
9. An example of nutrimillet
   (a) prosomillet (b) sorghum
   (c) maize (d) bajra
10. An early crop of rice is referred as
   (a) aus crop (b) aman crop
   (c) boro crop (d) zaid crop
11. An example of crop tolerant to alkaline soils
   (a) cotton (b) bajra
   (c) mustard (d) potato
12. One of the following is a drought tolerant crop
   (a) safflower (b) cabbage
   (c) sugarbeet (d) maize
13. The crop requires intercultivation
   (a) groundnut (b) fodder jowar
   (c) onion (d) paragrass
14. An example of a catch crop
   (a) greengram (b) cotton
   (c) tomato (d) redgram

15. A trap crop for striga weed
    (a) cowpea (b) mesta
    (c) safflower (d) mustard
16. A fouling crop
    (a) groundnut (b) cowpea
    (c) cotton (d) potato
17. Ladies finger is a
    (a) truck crop (b) cole crop
    (c) ware crop (d) cash crop
18. Napier grass is a
    (a) truck crop (b) soiling crop
    (c) riparian crop (d) ware crop

Chapter 4

Soil: Composition and Properties

The word soil is derived from the Latin word '*Solum*' means soil. In general the top 15–20 cm cultivated layer is referred as soil. It is a loose and unconsolidated outer layer of the earth crust that is crumbly and powdery in nature. Soil below the plough layer is called subsoil.

Soil is formed by weathering of rocks. Weathering may be physical or chemical or both. Factors of soil formation are parent material, climate, living things-plants, animals and microorganisms, topography and time. Soil formation is a function of parent material (P), climate (C), topography (T_o), life (L) and time (T) which is represented by f(P C To L T).

Importance of Soil

(a) Soil provides mechanical support to the plant: Roots anchored to the soil enable growing plant to remain erect.

(b) Nutrient supply: Soil supplies mineral nutrients 13 essential elements are required by the plants from the soil. Minerals undergo decomposition release nutrient elements.

(c) Nutrients are held on the colloidal surface due to presence of charge.

(d) Soil holds water. Water held in the soil is called soil moisture. It is held with cohesive and adhesive forces operating in capillary pores.

(e) Soil supplies oxygen to the roots.

(f) Retains heat and maintains the temperature required for the growth of roots.

Soil composition

Soil is composed of solid, liquid, gaseous and living phases:

Solid phase: materials having diameter less than 2 mm size constitutes soil. Solid phase comprises inorganic and organic constituents. Soils having more than 20% organic matter are designated as organic soils. Where inorganic constituents dominate is called mineral soil.

The solid mineral particles comprise about 45% of the soil volume and organic matter 5%. At optimum moisture for plant growth the pore space is divided roughly in half, 25% of the volume being water space and 25% of the volume by air. Texturally the mineral particles present in the soil are divided into sand (0.05–2.00 mm diameter) silt (0.002–0.05 mm diameter) and clays (<0.002 mm diameter). The minerals may be primary and secondary minerals (quartz, micas and feldspars, silicate clay and iron oxides) calcium carbonate (absent in soils having pH less than 7) soils having high content of calcium carbonate are called calcareous soils. Soluble salt content varies from 0.1 to 0.2% on dry soil weight basis. If the electrical conductivity of the soil is >4 dSm^{-1}, the soils are called saline soils. Such soils are harmful to the growth of plants. Soils having pH <7 contain free oxides of iron, aluminium and silica. Under strong acidic pH range, crop growth will be poor.

Organic constituents: are derived from the plant and animals. The leaf and litter of plants, excreta of animals and upon their death undergo decomposition by the action of microorganisms. The ultimate product of decomposition is humus. During the process of decomposition nutrients are released into the soil. This process is called mineralization. The humus and humus like fractions and other metabolic organic compounds of the soil phase constitute the soil organic matter. The organic matter content of most of the Indian soils range from 1-2%. In arid and semi-arid regions soil organic matter content is as low as 0.2% and in humid warm climate it may be as high as 10%.

Living organisms

Constitute soil macro and microflora. A gram of fertile soil contains billions of microorganisms. The living weight of the microorganisms may be about 4000 kg per ha and may constitute 0.01–0.4% of the total soil mass. Soil macroflora–roots of higher plants, microfauna: protozoa, nematodes; microflora-bacteria, fungi, actinomycetes and algae. Microorganisms bring about nitrification, ammonification, mineralization, phosphorus solubilizaion, mobilization of nutrients. Immobilization of nutrients. Some of the microorganisms are harmful to the plants. They cause diseases in crop plants, and loss of nutrients from the soil.

Liquid phase: The soil water is present in the soil pores. It is referred as soil solution. It is comprised of dissolved nutrients, salts and soluble organic constituents. The soil acts as a reservoir for supply of water to plants. There is often exchange of nutrient ions between soil solution and colloids, soil solution and plant roots.

Gaseous Phase: Soil air is rich in carbon dioxide (> 0.03%) and low in oxygen (5–10% as against 20 percent in atmosphere).Too high concentration of carbon dioxide in the soil cause toxic effects on plant roots. There should be free exchange of gases between soil and atmosphere to ensure free supply of oxygen to plant roots.

Soil Properties

Soil Texture and Structure

Physical properties of the soil like draught power requirement. Water holding capacity, water infiltration, soil erosion, nutrient supplying capacity, soil aeration, extent of root penetration is influenced by soil texture and structure.

Soil Texture: Refers to the size of the individual mineral particles and specifically refers to the relative proportions of various sized particles in a given soil. The size of particles in a mineral soil is not readily subject to change or it cannot be readily altered.

Textural classification of soil

< 0.002 mm	0.002–0.05 mm	0.05–0.01 mm	0.01–0.25 mm	0.25–0.50 mm	0.50–1.00 mm	1.00–2.00 mm	> 2.00 mm
Clay	Silt	Very fine Sand	Fine sand	Medium sand	Coarse sand	Very Coarse sand	Gravel

USDA system of textural classification

Soil textural classes

Sands: At least 70% sand 15% or less clay. Textural classes: Sands and loamy sands.

Silts: At least 80% silt 12 or less clay. Textural classes: Silt.

Clays: At least 35% clay separate, Textural classes: clay, sandy clay and silty clay.

Loams: Equal proportion of sand, silt and clay. If the proportion of sand silt and clay varies accordingly designated as–sandy loam, silt loams, silty clay loams, sandy clay loams and clay loams.

Structure of mineral soils

The term structure relates to the grouping or arrangement of the primary soil separates into secondary groupings called aggregates or peds.

Structural classes

1. **Spheroidal (Granular or crumb type):** If the granules are porous the term crumb is used. The peds are round or spherical. Generally found in surface layers, particularly rich in organic matter, and prominent in grasslands.
2. **Plate like or platy:** Aggregates arranged in relatively thin horrizontal plates. These are found in virgin soils and also in lower horizons.

3. **Prism like (Columnar and prismatic):** The aggregates are oriented vertically resembling pillars of varying heights with diameter of 15cm or more. These are found in subsurface horizons of poorly drained soils of arid and semiarid regions.
4. **Block like:** Blocks of six faces which are irregular found in sub soil.

   Structureless soil: Single grained structure as in sands.

   Massive soils–Loam or clay loams.

Genesis of granular and crumb structure

1. **Physical forces:** Root growth, alternate wetting and drying, freezing and thawing, activity of earthworms. The tillage operation at optimum soil moisture promotes formation of crumb structure.
2. **Binding agents:** Colloidal clay, gel and viscous materials (mucilage or mucus), humic and fulvic acids, polysaccharides, hemicelluloses and other natural polymers, humus adsorbed on clay surface referred as clay - humus complex, sesquioxides (Fe and A1 oxides) calcium and magnesium adsorbed on the soil particles act as flocculating agents.
3. **Soil conditioners:** Synthetic chemicals (Polymers) polyacrylamide (as low as at 0.00025%) concentration are able to aggregate soil particles.

   Structural management

   (a) Addition of organic matter.

   (b) Tillage at optimum soil moisture, fall ploughing.

   (c) Adoption of suitable crops and cropping system.

   (d) Avoid using irrigation water with sodium salts.

Significance of soil structure

If the soil has granular structure in the 0–20 cm depth it promotes pore space as a consequence there will be better soil aeration, root penetration, water movement, availability of plant nutrients, growth of microorganisms. Good soil structure decreases bulk density and moderates heat transfer.

Destruction of soil structure

Beating action of rain drops, repeated tillage with heavy implements, puddling the soil and lack of application of organic manures destroys the soil structure.

Porosity: Pore spaces in soil consists of that portion of soil volume not occupied by solids (either by mineral or organic). Pore spaces are occupied by air and water.

Macropores > 0.06 mm diameter

Micropores < 0.06 mm diameter

Pore space in different soils

In sandy soils	35–50%
Fine textured soils	40–60%
Compacted sub soils	25–30%

Significance

Proper proportion of macro and micro pores are essential to provide both aeration and water. If micropores are more soil will hold more water, aeration will be less. On the other hand if macropores are more there will be free movement of water, free drainage and less water holding capacity.

Management of soil porosity: Porosity is related to soil structure and soil texture. Those practices adopted to improve soil structure as well as good tillage practices will improve the soil porosity.

Particle density (PD): It is the mass per unit volume of soil solids. Expressed as Mega grams per cubic meter. It varies from 2.60–2.75 Mg/m^3 in most mineral soils while, in soils rich in organic matter will have 2.4 Mg/m^3.

Bulk density (BD): It is the mass of unit volume of dry soil. This includes both solids and pore space.
In clays, clay loams and silt loams the BD varies from 1.00 Mg/m^3 to 1.6 Mg/m^3.

In sands and sandy loams; it varies from 1.20–1.80 Mg/m^3.

The ideal BD for field crops in clays 1.4 Mg/m^3 while in sandy soils 1.6 Mg/m^3.

Friability: Characterizes the crumbling of soil mass. The moisture range in which the soils are friable (The soil mass can be easily crushed) is also the range in which conditions are optimum for tillage.

Questions

I. Answer the following

1. What are the constituents of soil?
2. What are the textural classes of soil?
3. What are the structural classes of soil?
4. What are the factors of soil structure formation?
5. What are the management practices for maintenance of good soil structure?
6. What are the causes for destruction of soil structure?

II. Define

(a) Particle density
(b) Bulk density
(c) Porosity
(d) Friability

III. Underline the correct answer

1. The organic matter content of mineral soils is
   (a) 5% (b) 45% (c) 25% (d) 20%
2. Texture of the soil is not related to
   (a) water holding capacity of the soil
   (b) nutrient supplying capacity of the soil
   (c) soil aeration
   (d) all the three stated under a, b and c
3. Clay soils contain
   (a) 70% clay (b) 35% clay
   (c) 15% clay (d) 80% clay
4. Structural class which is ideal for field crops
   (a) platy (b) spheroidal
   (c) prism (d) blocky
5. The cultural practice degrades soil structure
   (a) addition of agricultural lime
   (b) addition of organic manures
   (c) adoption of appropriate cropping system
   (d) puddling
6. Bulk density of soils ranges from
   (a) 1.0 to 1.6 Mg/m^3 (b) 1.6 to 1.75 Mg/m^3
   (c) 2.60 to 2.75 Mg/m^3 (d) 1.75–2.60 Mg/m^3

Chapter 5

Soil Fertility and Productivity

Soil fertility is the ability of the soil to supply all the essential nutrients in optimum amount in balanced proportion in a form readily available to plants under conditions favourable for plant growth and should be free from toxic substances.

Soil productivity: Soil productivity is defined as the capability of soil for producing a specified crop or sequence of crops under defined set of management practices. It is measured in terms of outputs or harvests in relation to the inputs of production factors for specific kinds of soils under a physically defined system of management according to USDA (1957).

Difference between soil fertility and productivity

Sl. No.	Soil fertility	Soil productivity
1.	Indicates the presence of nutrients	Capacity of the soil to produce crops
2.	Soil fertility can be improved by addition	Productivity of the soil is an inherent of nutrients property of the soil. It cannot be easily altered
3.	Can be assessed in the laboratory by testing the soil.	Can be assessed by growing the crop and measuring the yield per ha.

"All fertile soils are not productive"

Fertility is the measure of available nutrients present in the soil while, the productivity of the crops grown is not only influenced by the presence of available nutrients but other growth factors or limitation *viz.*, inadequate soil moisture, illdrained soils (Water stagnation), excess of salts, toxic levels of some essential elements of plant growth, extreme soil texture, shallow soils, presence of hard pans, soils infested with pathogens, parasitic weeds and problematic weeds.

Factors influencing the soil fertility

Natural factors

Physical factors

Soils formed from sand alone or quartz is poor in nutrients. Soils formed from apatite and feldspar is rich in phosphorus and potash respectively. Soils formed from calcite are rich in calcium while from dolomite are rich in calcium and magnesium. Soils formed from gneiss and schist is fertile. Soils formed from volcanic eruptions *viz.* Andesols are rich in nutrients.

Rainfall among the climatic elements plays a dominant role in altering the nutrients present in the soil. In arid and semiarid regions due to low rainfall, bases *viz.*, calcium, magnesium and sodium are accumulated in the surface layer while in humid regions due to high rainfall the bases are leached down and soils are rich in iron, manganese and aluminium. Besides in high rainfall areas organic matter production and additions more on the contrary, in arid and semiarid regions organic matter production and addition is poor. Besides, high temperature favours rapid oxidation of soil organic matter.

Topography alters the effect of parent material and rainfall. As topography becomes steeper surface soil along with nutrients is transported to the low lying areas along the direction of slope. As a consequence the soils in the upper elevation get depleted of nutrients while in the lower elevation and valleys where soil gets deposited becomes fertile.

Depth of the soil: Deeper soils are more fertile than shallow soils. The volume of soil available for root exploitation is more in deeper soils. Presence of hard pans or plough pans, clods restrict the availability of plant nutrients.

Type of clay minerals influence the soil fertility- If the soils are dominated by expanding type of clay minerals *i.e.*, 2:1 silica (tetrahedral): alumina or magnesia (octahedral) sheets *viz.*, smechtite, vermiculite or montmorillonite; or 2:1:1 (silica: magnesium: aluminium) *viz.*, chlorite hold more nutrients due high cation exchange capacity besides they can fix ammonium or potassium in the interlattice layers which is in dynamic equilibrium with soil solution ammonium or potassium.

Chemical factors

Presence of sesquioxides (iron and aluminium oxides) in acidic soils and calcium carbonate in saline/calcareous soils transforms soluble phosphate into insoluble forms of ferric or aluminium phosphates or tricalcium phosphate which is not immediately available to crop.

In salt affected soils due to presence of high concentration of divalent cations *viz.* Ca and Mg, and Na in alkaline soils interferes with the uptake of K. Micronutrients *viz.*, Zn, Fe, Mn and Cu become deficient as the pH increases (7.0–8.5)

Biological factors

Soil microflora and microfauna influence the soil fertility as they are involved in decomposition of soil organic matter, fixation of atmospheric nitrogen, involved in solubilisation and availability of various plant nutrients. Besides, macroflora and macrofauna *viz.*, Plant roots, earthworms, nematodes alter the soil fertility favourably by their exudations, size reduction of soil organic matter and through addition of their excreta improve the soil fertility.

Artificial factors influencing soil fertility

Addition of organic manures builds up soil fertility. Application of balanced fertilizers, optimum dose and time in conjunction with organic manures sustains the soil fertility. Adaption of appropriate cropping systems which add lot of residues to the soil, inclusion of legumes in rotation builds up soil fertility over a period of time.

Tillage at optimum moisture, good tilth, fall ploughing improves the availability of nutrients. Alley cropping, adoption of integrated farming system, agro forestry are the good agriculture practices to improve the soil fertility

Addition of lime to acidic soils and reclamation of salt affected soils through appropriate management practices improves the availability of plant nutrients.

Judicious irrigation management practices in irrigated situation and adaption appropriate soil moisture conservation practices and prevention of soil erosion sustains soil fertility.

Factors influencing soil productivity

Physical factors

- Soil texture- course textured soils are less productive than the fine textured soils,
- Water logged soils, ill drained soils, eroded soils, extreme soil temperature limit the soil productivity.
- Soil depth- Shallow soils are less productive than deep soils
- Topography- Soils in the higher elevation are less productive than in the lower region.

Chemical factors

- Presence of excess amount of salts, soil acidity, presence of toxic substances, toxic levels of essential plant nutrients, extremes of soil pH limits the soil productivity

Biological factors

- Infestation by problematic weeds, pathogens and pests hinders soil productivity.

What are good soil management practices?

Good soil management practices
• Good tilth • Control of weeds • Maintenance of adequate levels of organic matter • Adequate supply of plant nutrients • Control of pests and pathogens • Adoption of soil and water conservation practices • Adoption of suitable crop rotation • Providing drainage • Alleviating the soil from excess salts, acidity, toxic substances.

- Good tilth is the first feature of good soil management. It means a suitable physical condition of the soil and implies in addition, a satisfactory regulation of soil moisture and air.
- The maintenance of soil organic matter which encourages granulation is an important consideration of good tilth.
- Tillage operations and timings should be adjusted so as to cause the minimum destruction of soil aggregates. Good tilth minimizes erosion hazards.
- The choice and sequence of adaptable crops or crop rotation are other very important considerations. These are related to climate, particularly rainfall and its pattern of distribution and the characteristics of the soil profile, including drainage and extent and duration of available soil moisture. A proper sequence of crop varieties greatly influences soil conditions. It is more realistic to evolve cropping patterns and land management practices according to land capability. Cropping patterns chosen and management practices adopted should aim at soil and moisture conservation for efficient nutrient and moisture utilization.
- In irrigated areas, special management practices become necessary to avoid salinity, alkalinity, water logging, leaching and the loss of plant nutrients. In rainfed areas special management practices include improving soil conditions to receive, retain and release more soil moisture, harvesting water to use as life saving irrigation or extending the cropping season when there is insufficient rainfall for raising crops, protecting the soil from degradation both in cropped and bare fields. Land shaping and levelling, mulching and the use of wind brakes and vegetative cover are the other major aspects.
- The productive capacity of the soil should never be allowed to diminish, but rather should be improved and maintained by providing adequate organic manures and plant nutrients through fertilizers and by including legumes in the rotation and the use of biofertilizers. Similarly the provision of irrigation

facilities in semi arid and arid areas, adoption of different remedial measures against excessive salinity and alkalinity or acidity in humid areas, use of specific soil amendments to correct imbalances of plant nutrients and application of micronutrients where they are deficient.

- Economic plant protection measures against pests, pathogens and parasites including weeds should form part of the management practices in the cropping system. This can be achieved by following recommended cultural practices or by application of pesticides/fungicide/herbicides.
- The management practices adopted should be economically profitable and emphasis should be laid on maximizing sustained income rather than yields for the time being. An integrated land plan including all the above points and economically profitable should be developed for individual situations.

Results of bad soil management

- Soil erosion : tillage practices should be oriented to conserve soil and water.
- Soil exhaustion: avoid continuous cultivation of exhaustive crops, regular use of organic manures and fertilizers.
- Salt accumulation: poor drainage, use of poor quality water, toxic substances from synthetic fertilizers and agro industrial wastes.
- Infestation with perennial weeds.
- Structure of the soil is spoiled.
- Lower productivity and profits.

Questions

I. Define

1. Soil fertility
2. Soil productivity

II. Answer the following

1. Justify the statement "All fertile soils are not productive".
2. What are the physical factors influencing soil fertility?
3. What are the chemical and biological factors influencing soil fertility?
4. What are the artificial factors influencing soil fertility?
5. What are the factors of soil productivity?
6. What are the good soil management practices?
7. What are the results of bad soil management?

III. Underline the correct answer

1. One of the following is not related to the soil fertility
   (a) presence of nutrients
   (b) it is dynamic
   (c) can be assessed by testing the soil in the laboratory
   (d) assessed by the yield or harvests per unit area
2. Physical factor which does not influence the soil fertility
   (a) Soil texture
   (b) rainfall
   (c) topography
   (d) soil colour
3. Chemical factor related to the soil fertility
   (a) presence of sesquioxides
   (b) concentration of carbonates and bicarbonates of calcium and magnesium in soil
   (c) CEC of the soil
   (d) all the three under a,b and c.
4. These soils are less productive
   (a) fine textured soils
   (b) well drained soils
   (c) deep soils
   (d) soils of steep slopy lands
5. These soils are less productive
   (a) soils having extreme pH
   (b) soils with toxic substances
   (c) soils having toxic levels of plant nutrients
   (d) all the three under a, b and c
6. Good soil management practices are
   (a) good tilth
   (b) maintenance of adequate soil organic matter
   (c) minimize loss of soil and water
   (d) all the three under (a), (b) and (c)
7. Results of bad soil management
   (a) soil erosion
   (b) soil exhaustion
   (c) salt accumulation
   (d) all the three stated under (a), (b) and (c)

Chapter 6

Tillage and Tilth

Tillage is the physical manipulation of the soil with tools and implements which, results in good tilth for better germination and subsequent growth of crops.

Tilth is the physical condition of the soil resulting from tillage. Soil is said to be in good tilth when it is mellow, friable and adequately aerated. Tilth is dynamic. Mechanical forces may change the roughness of the soil surface, the total porosity and bulk density of the tilled layer and the aggregate or clod size. Heavy rainfall and high velocity wind or water destroys the tilth. The action of wetting and drying, freezing and thawing regenerates desirable tilth. The roughness of the soil surface is an index of the amount of water that can be stored in soil depressions and may also be related to the resistance of the soil surface to sealing. Roughness may be related to rate of evaporation, transfer of heat and air between the soil and atmosphere.

Objectives of tillage

The primary purpose of tillage is often to reduce the aggregate or clod size. Aggregates must be small enough to achieve good contact between the roots of the seedlings and soil in order to prevent drying of the soil, to provide sufficient soil solution, aeration. Yet the aggregates should not be so finely divided as to encourage severe crusting when dry.

- To provide suitable seed bed for sowing the seeds necessary for germination and emergence of seedling or for transplanting seedlings or planting materials. Fine seed bed is desirable for smaller sized seeds while rough seed bed is enough for larger seeds.
- To control of weeds, pests and diseases
- The residues of the previous crop are incorporated into the soil.
- To improve the physical condition of the soil *viz.* structure (development of granular structure), porosity, aeration, water holding capacity, infiltration and reduction in bulk density and run off. The hard pans are broken during ploughing which facilitate greater infiltration and root penetration into deeper layers. The positive influence of tillage on these properties will ultimately reduce soil erosion.

- To improve the nutrient availability by physical process.
- To hasten the chemical and biological processes in the soil. This in turn has impact on the activity of microorganisms and organic matter decomposition.

Influence of tillage on soil physical properties, temperature, soil moisture and root penetration

Soil structure, porosity and bulk density: Frequent traffic due to use of heavy implements and tractor/ heavy machinery, repeated tillage operations, puddling, tillage in hard soil/ at low moisture condition destructs soil structure. Tillage at optimum soil moisture improves the soil structure. Soil porosity increases with tillage in the plough layer while the soil layer below the plough sole gets compacted and porosity decreases. The soil bulk density decreases in the plough layer.

Soil temperature: Tillage influences soil thermal properties. Loosening the soil decreases the heat capacity because of larger volume fraction of soil air. Thermal conductivity is decreased with decreasing fraction of solids and increased with increasing water content. Loosening the soil often results in a larger difference between maximum and minimum temperatures but this may be comparatively less than air temperature. Soil temperature affects germination, nutrient availability, plant development, grain yield and nutrient content of grains.

Soil moisture: Tillage may have marked influence on soil moisture through its effect on infiltration, run off, temporary surface storage, internal storage and availability to plants. Roughness of soil surface is an index of the amount of water that can be stored in depressions and may also related to resistance to sealing. Tillage creates uneven micro-relief which can store considerable water in the small depressions for later infiltration. Once water enters the soil its rate of movement will depend on the internal transmission characteristics of the profile. Increased porosity from tillage may act as an important reservoir for temporary storage of water during rains. The amount water retained by a soil may be influenced by the soil density and aggregate size. Ploughing or loosening a dense soil provides pores which can store water and may increase the storage capacity. In loose soils when surface soil is dry water vapor moves to surface. If there is rainfall, infiltration is higher. In compact soil at relatively higher moisture content there will be greater conductance of water to the evaporating surface.

Soil aeration: Soil aeration is the mechanism of exchange of oxygen and carbon dioxide between soil pore space and the atmosphere in order to prevent the deficiency of oxygen and toxicity of carbon dioxide. Respiration by plant roots and microorganisms depletes oxygen and releases carbon dioxide and minute quantities of other gases into the soil atmosphere. The amount of carbon dioxide varies widely depending on temperature, organic matter and microbial respiration. A constant inflow of oxygen and out flow of carbon dioxide is essential for plant growth. Loosening a dense soil facilitates exchange of gases between soil air and atmosphere. Air filled pores commonly fluctuate from 15 to 30 percent of the total volume depending upon water content, soil density and structure.

Causes of poor soil aeration

- Compact soil
- Water logging
- Soils having excessive amount of readily decomposable organic matter

Harmful effects of poor aeration

- Microbial activity is reduced, slow decomposition of organic matter
- Rhizobium cannot symbiotically fix atmospheric nitrogen
- Abnormal development of roots *e.g.*, sugarbeet and carrot
- Anaerobic bacteria decompose soil organic matter in the complete absence of oxygen release toxic substances like sulphides which are harmful to crop plants

Root penetration: plant roots can exert tremendous forces in penetrating the soil. In packed soils pore size may be too small for root tips to enter. Critical pore size varies with kind of plant, but there are pore diameters below which roots may not penetrate. Pore rigidity as well as size will affect the root penetration.

Types of tillage

Tillage operations may be grouped into

1. On season tillage
2. Off-season tillage

On-season tillage

Tillage operations that are done for raising crops in the same season or at the onset of the crop season are known as on season tillage. They may be preparatory tillage and inter tillage.

Preparatory tillage

This refers to tillage operations that are done to prepare the field for raising crops. It consists of deep opening and loosening of the soil to bring about a desirable tilth as well as to incorporate or uproot weeds and crop stubble when the soil is in a workable condition.

Types of preparatory tillage

1. Primary tillage
2. Secondary tillage

Primary tillage

The tillage operation that is done after the harvest of crop to bring the land under cultivation is known as primary tillage. Ploughing is the opening of compact soil with the help of different ploughs *viz*., country plough, mould board plough, bose plough. Bullocks, tractors and power tillers are used to draw the primary tillage implements.

Secondary tillage

The tillage operations that are performed on the soil after primary tillage to bring a good soil tilth are known as secondary tillage. Secondary tillage consists of lighter or finer operation which is done to clean the soil, break the clods and incorporate the manure and fertilizers. Harrowing and planking is done to serve those purposes. Planking is done to crush the hard clods, level the soil surface and to compact the soil lightly. Harrows, cultivators, guntakas and spade are used for secondary tillage.

Tertiary tillage/leveling

Leveling board, buck scrapers *etc*. are used for leveling

After tillage (Inter tillage)

The tillage operations that are done in the standing crop after the sowing or planting and prior to the harvesting of the crop plants are called after tillage. This is also called as inter cultivation or post seeding planting cultivation. It includes harrowing, hoeing, weeding, earthing up, drilling or side dressing of fertilizers *etc*., Spade, hoe, weeders *etc*., are used for inter cultivation.

Dry tillage

Dry tillage is practiced for crops that are sown or planted in dry land condition having sufficient moisture for germination of seeds. This is suitable for crops like broadcasted paddy, jute, wheat, oilseed crops, pulses, potato and vegetable crops. Dry tillage is done in a soil having sufficient moisture (21–23%). The soil becomes more porous and soft due to dry tillage. Besides this the water holding capacity of the soil and aeration are increased. These conditions are more favourable for soil micro-organisms.

Wet or puddling tillage

The tillage operation that is done in a land with standing water is called wet or puddling tillage. Puddling operation consists of ploughing repeatedly in standing water until the soil becomes soft and muddy. Puddling creates an impervious layer below the surface to reduce deep percolation losses of water and to provide soft seed bed for planting rice. Puddling tillage is done in both the directions for the incorporation of green manners and weeds. Wet tillage (puddling) destroys the soil structure and the soil particles that are separated during puddling settle later. Wet tillage is the only means of land preparation for transplanting semi aquatic crop plant such as rice. Planking after wet tillage makes the soil level and compact. Puddling hastens transplanting operation easily and smoothly as well as establishment of seedlings. Wet land ploughs or worn out dry land ploughs are normally used for wet tillage.

Off-season tillage

Tillage operations done for conditioning the soil suitably for the forthcoming main season crop are called off-season tillage. Off season tillage may be

1. Post-harvest tillage
2. Summer tillage
3. Winter tillage
4. Fallow tillage

Special purpose tillage

Tillage operations intended to serve special purposes are said to be special purpose tillage. They are:

1. sub soiling
2. leveling
3. blind tillage
4. clean tillage
5. wet tillage and
6. zero tillage.

Sub soiling

To break the hard pan beneath the plough layer special tillage operations (chiseling) are performed to reduce compaction. Sub soiling is essential once in four to five years where heavy machineries are used for field operations, seeding, harvesting, transporting *etc.*

Advantages

- Greater volume of soil may be obtained for cultivation of crops.
- Excess water may percolate downward to recharge the permanent water table.
- Reduce runoff and soil erosion.
- Roots of crop plants can penetrate deeper to extract moisture from the water table.

Clean tillage is one wherein 100 percent of the top soil is mixed or inverted. With clean tillage all the plant residues are removed and buried. The growth of weeds is prevented.

Conventional tillage/Traditional tillage: Conventional tillage has been defined as combined primary and secondary tillage operations performed in preparing the seed bed. Generally there exists narrow difference between clean and conventional tillage.

Suitability

- Adopted in Class-I lands

Advantages

- Weeds are efficiently controlled
- Crop residues are thoroughly incorporated
- Better microbial activity

Disadvantages

- More energy requirement
- Greater loss of soil moisture
- Formation of hard pan
- Surface soil is more prone to erosion

Wet or puddling tillage

The tillage operation that is done in a land with standing water is called wet or puddling tillage. Puddling operation consists of ploughing repeatedly in standing water until the soil becomes soft and muddy. Puddling creates an impervious layer below the surface to reduce deep percolation losses of water and to provide soft seed bed for planting rice.

Depth of ploughing

The desirable depth of ploughing is 12.5–20.0 cm. The ploughing depth varies with effective root zone of the crop. The depth of ploughing is 10–20 cm for shallow rooted crops and 15–30 cm for deep rooted crops.

Number of ploughing

Number of ploughing depends on soil conditions, time available for cultivation between two crops, type of cropping systems. Zero tillage is practiced in rice fallow pulses. Minimum number of ploughing are taken up at optimum moisture level to bring favourable tilth depending on need of the crop.

Time of ploughing

The optimum soil moisture content for tillage is 60% of field capacity.

Requirements of ideal seed bed

An ideal seed bed is one which sustains all stages of crop growth and development starting from germination and emergence to maturity without great deterioration and depletion.

1. Seed bed should be free from large clods, crop residues and established weeds.
2. Should be properly levelled and of a desirable physical state with the correct moisture content for good growth and yield of crop.
3. There must be provision for adequate irrigation channels and drainage.
4. The soil should be supplied with basal dose of manures and fertilizers and thoroughly mixed into the soil.
5. Stiff stubbles, stalks and other not readily decomposable organic matter should be removed otherwise they may invite infestation of pests and pathogens.
6. Granulation should be retained. Avoid excessive tillage than the required. Poor granulation encourages soil erosion.
7. Depth of tillage should be need based.
8. There must be adequate moisture and air supply in the seed bed.
9. Rough seed beds are favoured for winter and large seeded crops, firm and shallow seedbed for fine seeded crop. For soil moisture conservation fine seed beds during summer in light soils is essential.
10. Unweathered cloddy soil should not be brought to the surface by deep tillage.
11. Loose granular seed bed for drylands while puddle seed bed for wet lands.
12. Seed beds should be prepared quickly to avoid loss of moisture. Soil moisture loss can be prevented by breaking the continuity of the capillary pores, sealing of pores and also by reducing exposed area by levelling, planking or harrowing.
13. Proper land shaping and configuration is essential for soil moisture conservation *e.g.*, ridge and furrow, broad bed and furrow or flat bed depending upon the situation and crop requirement.

Don’ts during seedbed preparation

- Seedbed preparation should not be prolonged.
- Avoid excessive tillage.
- Tillage should be carriedout at optimum soil moisture.

Modern concepts of tillage

- Minimum tillage
- Zero tillage
- Conservation tillage
- Stubble mulch tillage
- Blind tillage

Minimum tillage may be defined as a group of land preparation methods for planting in which the number of tillage operations over the field is less than conventional tillage. This can be achieved by omitting the tillage operations which do not give much benefit when compared to the cost. For example combining seeding and fertilizer application, row zone tillage, plough plant tillage, wheel track planting.

Suitability

Medium textured soils. Minimum tillage is practiced in black soil to minimize water loss for establishment of chickpea, safflower and rabi jowar.

Advantages

- Reduced soil compaction
- Better soil conservation
- Energy requirement is less
- Reduced labour and machinery
- Time saving

Disadvantages

- Low seed germination and establishment. Difficulty in sowing
- Nodulation is adversely affected
- Use of herbicides is indispensable
- Decomposition of organic matter is slow
- Perennial weeds may become dominant

Zero tillage is an extreme form of minimum tillage. Primary tillage is completely avoided and the secondary tillage is restricted to seedbed preparation in the row zone only. Planting is done in untilled land by opening a narrow slot or trench or band only of required width and depth for sowing and covering the seed or planting the seedling.

Weeds are taken care by using broad spectrum nonselective and non-persistent herbicide before sowing, subsequent to sowing by using selective and persistent herbicides.

Till planting is one of the zero tillage method in which heavy machineries are used to clean a narrow strip over the crop row. Then a narrow band of soil is opened. Seeds are placed and covered.

Advantages

- Saving in energy, labour and time
- Reduced compaction
- Increased earthworm activity and soil organic matter. Suitable for organic farming

Disadvantages

- Difficult to establish optimum crop stand
- Mineralization of soil organic matter is slow
- Nitrogen requirement of crops is high
- Build up of perennial weeds and pests

Conservation tillage is the tillage operation performed to reduce soil erosion and to conserve soil moisture is referred as conservation tillage. This is achieved by covering the soil at least by 30 percent of the land surface by the crop residue. Conservation tillage operations include reduced tillage operations like minimum tillage, no tillage, mulch tillage.

Stubble mulch tillage is method of tillage in which a mulch crop is grown during the fallow period or the stubbles of the previous crop are uprooted and brought to the surface and spread during tillage operation. The main objective is to protect the soil from erosion. Suitable in sloppy areas and drylands.

Advantages

- Reduced loss of soil and water
- Improved organic matter content

Disadvantages

- Less effective weed control
- Stubbles interfere in sowing operation

Blind tillage is the tillage of the soil after sowing a crop either before the crop plants emerges or while they are in early stages of growth. It is extensively employed in sorghum and drilled paddy where emergence of crop seedlings is hindered by soil crust formation on receipt of rain or by irrigation immediately after sowing. Shallow harrowing with entire blade harrow without disturbing the emerging crop seedlings will loosen the soil crust and help in emergence of seedlings. Generally weed seedlings emerge within two to three days after sowing while many cereals take 7 to 8 days for seedling emergence. By blind tillage weeds are killed at their early stages.

Questions

I. Answer

1. What are the objectives of tillage?
2. What are the characteristics of ideal seed bed?
3. What are the modern concepts of tillage?
4. What is blind tillage?
5. What is conservation tillage?
6. What is the tilth?
7. Describe the effect of tillage on
   (a) soil structure (b) soil porosity
   (c) soil temperature (d) soil moisture
   (e) soil aeration (f) root penetration

II. Define

1. Primary tillage
2. Secondary tillage
3. After tillage
4. Puddling
5. Off season tillage
6. Conventional tillage

III. Underline the correct answer

1. One which is not related to the beneficial effects of tillage
   (a) providing ideal seed bed
   (b) control of weeds
   (c) increase bulk density of soil
   (d) improves the structure of the soil
2. Tillage
   (a) increases porosity
   (b) decreases bulk density
   (c) at optimum soil moisture improves soil structure
   (d) increases the particle density
3. Tillage
   (a) increases infiltration of water
   (b) reduces runoff of water
   (c) improves soil moisture retention
   (d) all the three stated under a, b and c

4. Soil air is rich in
   (a) carbon dioxide
   (b) oxygen
   (c) nitrogen
   (d) carbon dioxide and oxygen
5. Poor soil aeration causes
   (a) decline in microbial activity
   (b) accumulation of toxic gases
   (c) decline in nodulation and nitrogen fixation
   (d) all the three under (a), (b) and (c)
6. Root penetration is related to the
   (a) pore size and pore rigidity
   (b) depth of sowing the seeds
   (c) texture of the soil
   (d) land topography
7. Ploughing is a
   (a) primary tillage
   (b) secondary tillage
   (c) after tillage
   (d) tertiary tillage
8. Hoeing is
   (a) after tillage
   (b) tertiary tillage
   (c) secondary tillage
   (d) primary tillage
9. Puddling reduces
   (a) soil erosion
   (b) infiltration of water
   (c) run off of water
   (d) evaporation of water
10. Summer tillage is a
    (a) off season tillage
    (b) special purpose tillage
    (c) after tillage
    (d) clean tillage
11. Clean tillage
    (a) 100 percent plant residues are removed from the surface
    (b) 50 percent plant residues are removed from the surface
    (c) 30 percent plant residues are removed from the surface
    (d)15 percent plant residues are removed from the surface
12. Depth of ploughing for shallow rooted crops
    (a) 10–20 cm
    (b) 15–30 cm
    (c) 5–10 cm
    (d) 30–45 cm
13. Optimum soil moisture for ploughing
    (a) 15 percent of field capacity
    (b) 30 percent of field capacity
    (c) 60 percent of field capacity
    (d) 90 percent of field capacity

14. Use of broadspectrum non selective herbicide is indispensible in
    (a) minimum tillage (b) zero tillage
    (c) conservation tillage (d) conventional tillage
15. Tillage of soil after sowing a crop but before emergence is called
    (a) secondary tillage (b) intertillage
    (c) blind tillage (d) clean tillge
16. Thirty percent crop residues of the previous crop is left on the land surface under
    (a) conservation tillage (b) stubble mulch tillage
    (c) minimum tillage (d) zero tillage

Chapter 7

Seeds and Sowing

A seed consists of a plant embryo, stored food material surrounded by a seed coat. Seeds of cereals and grasses consist of embryo or germ, endosperm and the pericarp.

Seeds of legumes contain no endosperm, the bulk of the seed being made up of the two cotyledons. The cotyledons constitute the food storage portion.

The embryo of the grass seed is made of epicotyl, hypocotyl and scutellum- the single cotyledon of monocots, the epicotyl gives rise to the shoot and the hypocotyl produces the roots.

Planting materials

Plants are propagated by asexual/ vegetative method and sexual method.

Vegetative method: parts of the plant other than the seed is used

1. Apomictic embryos: development of an embryo from cell other than a fertilized egg. *e.g.*, citrus
2. By runners or stolons: long slender side branches grow above ground. At each node shoots and roots develop *e.g.*, straw berry, white clover
3. Layerings: apple, pomegranate
4. Sucker: is a laterally growing subterranean off shoot from the base of the main stem of a plant *e.g.*, cardamom, banana
5. By separation:

   Bulbs: onion, garlic, cloves

   Corms: swollen base of a stem axis with distinct nodes and internodes *e.g.*, ginger, turmeric, gladiolus
6. By division: specialized stem structures *e.g.*, pineapple, canna
7. Offsets: pseudobulb is long and jointed with many nodes. Offsets are developed at these nodes. Roots develop from the base.
8. Tubers: potato
9. Tuberous roots: sweet potato
10. Crowns: straw berry
11. Cuttings: red raspberry, lemon

12. Stem cutting: rose, sugarcane, drum stick
13. Rooted slips: basal two are three internodes of the stem with a few roots *e.g.*, paragrass, guinea grass, hybrid napier
14. Leaf cutting: bignonia
15. Grafting : mango
16. Budding: rose, mango, grape
17. Micro propagation:
    Meristem cultures: orchids
    Tissue culture: tobacco, potato, banana

Qualities of good seed

Seed is the basic tool and is the principal means to get higher yields.

- Seed should be distinctly superior in quality
- Varietal purity
- Freedom from admixtures
- High germination percentage
- High vigour
- Healthy without infection
- Optimum moisture content

Use fresh seeds, quality seeds that facilitate higher germination and early establishment. Seeds, tubers, bulbs and setts are used as seed material.

Seed treatment

Seed treatment helps to prevent carriage of insects or disease causing pathogens in, on or with the seed. It also enables the seed to overcome seedling infection by soil borne fungi. Seed treatment serves as disinfectant when the pathogen on the seed is destroyed or inactivated or as a protectant when fresh entry of the pathogen is prevented.

Seed treatment is one single operation which returns more money for the amount invested.

Types of seed treatment

1. **Seed treatment for convenience of sowing**

   Cotton fuzz causes inconvenience in sowing causes clinging of seeds difficult to sow seeds either through seed drill or by broadcasting. Seeds are treated with cowdung slurry or concentrated sulphuric acid for two minutes.

2. **Mechanical treatment**

   The seeds of coriander are prepared by splitting them into two locules of which the fruit is made up, so the seeds fall apart and both economy and evenness of sowing are ensured.

   Garlic - bulb is split into individual cloves for sowing.

3. **Seed dilution**

   Sand is mixed with seeds like ragi, sesamum, tobacco. It is diluting the bulk so as to obtain a larger quantity which can be handled conveniently.

II. Methods to prevent the insects or pathogens

1. **Dust treatment**: Bulk of the seed is treated with dust formulations using seed drums.

   e.g., Captan 75% dust, Thiram 75% dust, Agroson 1% dust
2. **Wet treatment:** Chemicals are dissolved in water and the seeds/ planting material are dipped for specific period and dried *e.g.*, 4g mancozeb/ lit or 3g capton/ lit or 2g mercuric chloride/lit water used for treating potato seed tuber pieces. Cerasan wet at 2.5% is used for other crops.
3. **Slurry treatment** is suitable for applying wettable powder(WP) formulations *e.g.*, Captan, Thiram 75% WP.
4. **Pelleting:** It is used mostly as protectant against soil organisms and as a repellent against birds and rodents *e.g.*, $CaCO_3$ or lime is used to pellet the seeds.

III. Treatment with cultures

Seed coating: is the addition of materials to seed which may aid in growth and survival of seedling or plants. *e.g.*, Rhizobium cultures for legume seeds to increase nitrogen fixation.

IV. Seed treatment to break dormancy

Scarification: The dormancy due to hard seed coat or impermeable seed coats can be broken by scarification of seed coats (Rupture of seed coats). Seed coat in this case is impermeable to water and oxygen. Scarification should be done in such a way that the embryo should not be injured by chemical scarification. Mineral acids like HNO_3, HCl, H_2SO_4 are used for scarification at 0.1 to 0.5% for few minutes. 0.5% succinic acid or acetone or alcohols or NaOH also used.

Temperature treatment stratification (low temp pre-treatment is called stratification)

The dormancy due to physiological changes in embryo or due to after ripening may be broken by different temperature treatments. The temperature treatment hastens physiological changes that result in after ripening. This is mostly applicable when dormancy is inherent in the embryo. Low temperature treatment of 2–8°C for 12–24 hrs for pre-soaked seeds of 36 hrs.

Exposure of seeds to light

The light has an effect on hastening the germination of seeds of some species *e.g., Vernonica longifolia* seeds are exposed to the red light for 1–2 hrs at 10-15°C for 24 hr soaked seeds.

Tobacco seeds exposed to red light for 64 hrs enhance germination.

Application of pressure to seeds

Germination of Madicago sativa is found to be increased from 50 to 200%, when a hydraulic pressure of 2000 atm. at 180°C is applied. The reason may be, increase in permeability to water and oxygen.

Seed soaking

(a) Sunflower seeds are soaked overnight in water.

(b) Seed hardening: Ragi seeds are soaked for 18 hrs in water (600 ml water/kg of seed). In the next morning the first sign of germination is seen. Then the seeds are dried in shade for 24 hrs.

Treatment with micro nutrients: seeds are also treated with micronutrients in different ways.

Ragi: Prepare a paste of 12.5 g of $ZnSO_4$ with 4-6 spoonful of water in an earthen container mix with 2 to 2½ kg seeds, keep the contents over night and sow on the next day. Ammonium molybdate is coated at 2 g per kg seeds of redgram. Similar treatment is followed in soybean and groundnut.

Methods of establishing the crop

(a) Direct sowing

(b) Transplanting

Methods of sowing

1. Broadcasting
2. Line sowing /sowing behind the plough
3. Drill sowing
4. Dibbling
5. Planting

Broadcasting

Seeds are spread uniformly over well prepared land and are covered by ploughing or planking.

Disadvantages

1. The seeds fall at different depths when broadcasted resulting in uneven stand. The plants are at different height, size and age.

2. There is lot of wastage of seeds. Seeds fallen on the surface or >10 cm depth may not germinate, birds may pick up seeds.
3. No possibility of intercultural operations to control weeds.
4. The spacing available for individual plants vary considerably. Excess competition at certain areas and no competition in some areas of the field.

This method is used for sowing fodder crops or crops where seeds are cheap or crops which can easily establish and suppress weeds *e.g.*, paddy, ragi, wheat and horse gram.

Line sowing: Using a country plough and draw tube the seeds are dropped into the furrow directly.

Drilling: Crops can be sown in lines by using seed drills.

Advantages

- Weeds can be controlled economically by inter cultivation in line sown crops
- Facilitates uniform depth of sowing resulting in uniform crop stand.
- Seed rate can be considerably reduced by drilling.

Planting: Generally crops established by vegetative methods *e.g.*, potato sugarcane, hybrid napier *etc.* using tubers, setts, rooted slips *etc.* is referred as planting.

Dibbling: When individual seeds are placed manually at proper depth and the space is regulated is called dibbling *e.g.*, maize, cotton. There is no wastage of seed in this method.

Types of sowing

Dry sowing: Field is prepared with summer rains and seeds are sown in dry soil around 7 to 10 days before the anticipated receipt of sowing rains. The seeds germinate after the receipt of rains. Adopted in black soil where land preparation is difficult after *Kharif* rains.

Wet sowing: Two or three days after soaking rain, sowing can be taken and continued for two to three days. The minimum amount of rainfall necessary for taking up sowing is 20 mm.

Depth of sowing of seeds depends on

- Size of the seed
- Seed reserve
- Coleoptile length
- Soil moisture

Crops with bigger sized seed *e.g.*, groundnut, carrot, sunflower can be sown even upto a depth of 6 cm. Small sized seeds like tobacco, sesamum, ragi have to be sown as shallow as possible. The surface layer dries up quickly. So when small seeds are sown it should be watered frequently. If the seeds are sown deep the food reserves are exhausted to meet the long coleoptile growth even if the seedling emerges it will not be vigorous. Traditional wheat has long coleoptiles, while Mexican wheat has shorter coleoptile. Hence the Mexican wheat should be sown at shallow depth. Optimum depth of sowing for most of the field crops is 3 to 5 cm. Depth of sowing must not be greater than 20 times the seed size.

Problems of seedling establishment

- Crust formation: In soils which slake when wet and become hard when dry.
- Depth of sowing/shallow or deep.
- Degree of compactness.
- Quality of seeds.

After sowing practices to maintain optimum plant population

Thinning: removal of excess seedlings where plant population is thick. Carried out at 10–12 days of sowing.

Gap filling: wherever the seedlings have not established the seedlings grown in the nursery or polybags or the seedlings removed from the field where there is excess population are planted to maintain optimum plant population is known as gap filling.

Transplanting: The seedlings are raised in the nursery and after certain age, they are pulled out and planted in the main field.

Nursery

The seeds are first sown in small area called nursery, all the care is taken to raise the seedlings. When the seedlings attain certain age they are pulled out and transplanted in the main field.

Advantages

1. Small seeded crops *viz.*, tobacco (0.08–0.09 mg) chillies, tomato.
2. For better weed control - paddy, ragi
3. To save resources - water in paddy
4. For uniform crop establishment
5. Better care and maintenance in the early stages of crop growth.
6. Save seed material.

Disadvantages

1. More labour is required.
2. Under hot sun seedlings are killed if adequate irrigation is not available.

Nursery techniques: the points to be considered are

1. Choice of the nursery field/site
2. Land preparation
3. Adequate irrigation, drainage
4. Manuring, pest and disease control
5. Exposer to sunlight

Care should be taken to produce healthy and vigorous seedling.

Nursery site selection

1. Fertile soil
2. Good drainage
3. Avoid shade, under trees or borders
4. Adequate irrigation
5. Soil should be free from perennial weeds, pathogens and pests.
6. Should be near the main field where transplanting is taken
7. Avoid site where the nursery is raised during previous years to prevent contamination of varieties and carryover of diseases.

Nursery methods

1. Dry nursery
   (a) Raised beds
   (b) Flat beds
2. Wet nursery -Followed for paddy crop

Good quality seedlings

1. Vigorous
2. Good root development
3. Free from pest and disease incidence

Optimum age of seedlings for transplanting

1. Conventional paddy - 21 days to 35 days
2. SRI method - 13 days to 15 days
3. Finger millet - 17 days to 20 days
4. Chilli - 30 days to 45 days
5. Tobacco - 49 days to 56 days (or pencil thick seedlings)

Crop Density and Geometry

Crop density is defined as the number of plants per unit area. Maize planted at 60 cm × 20 cm will have 83,333 plants/ha while at 60 cm × 30 cm will have 55,555 plants/ha.

Planting geometry: It is the pattern of distribution of plants over the ground or shape of the area available for individual plants.

Square planting: If the inter row and intra row spacing is equal and the lines are arranged perpendicular to each other. Followed in wide spaced crops *e.g.*, Tobacco, Chilli, Cotton.

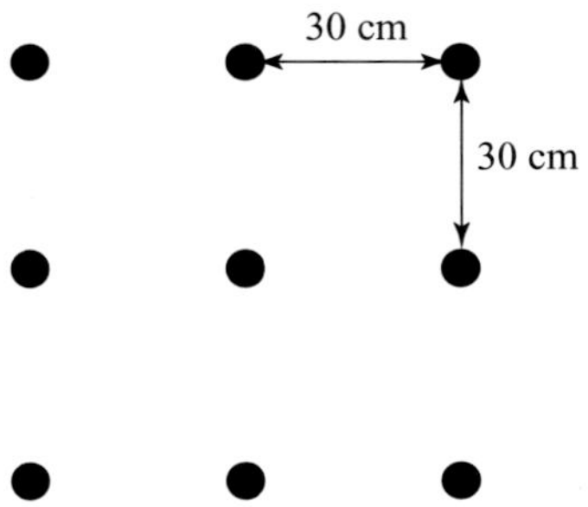

Advantages

1. Interculture operations can be done in both the direction.
2. All the plants get uniform space .

Rectangular planting–Inter row spacing is greater than intra row spacing.

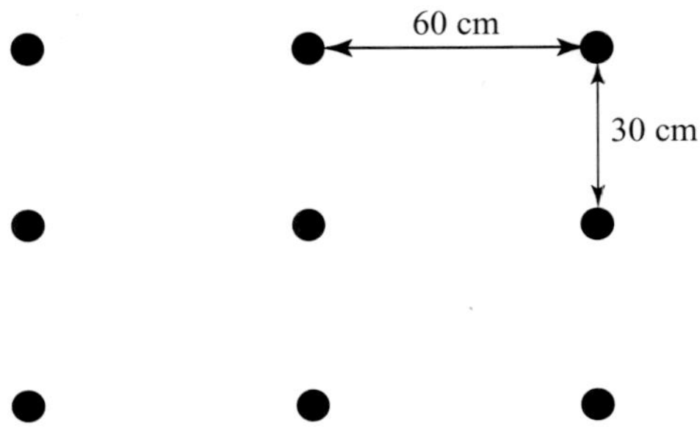

Inter-cultivation possible in one direction.

But the rows are arranged perpendicular to each other.

Triangular planting

Seeds or seedlings are planted in rows at a definite spacing. Intra row spacing may not be uniform. The plants are not aligned at right angle to each other.

Most of the crops sown using seed drill *e.g.*, sorghum, redgram,

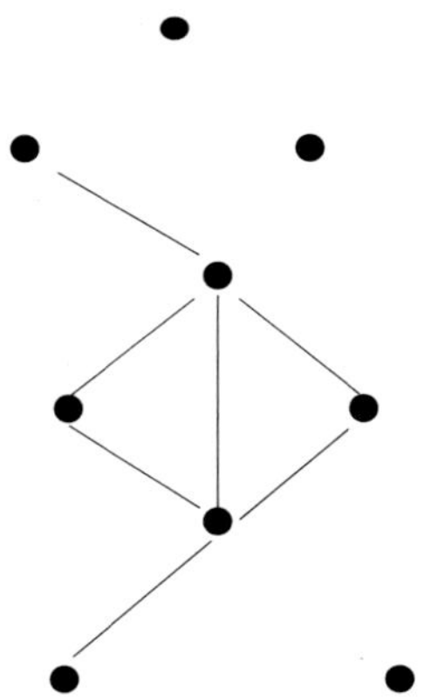

Optimum plant density: Is the number of plants per unit urea which utilize efficiently available growth factors *viz.* water, nutrient, light and carbon dioxide to produce maximum yields.

Normally maximum yields are obtained from plant population which does not allow plants to achieve their individual maximum potential. The entire community of plants is considered for higher production than the individual plant performance.

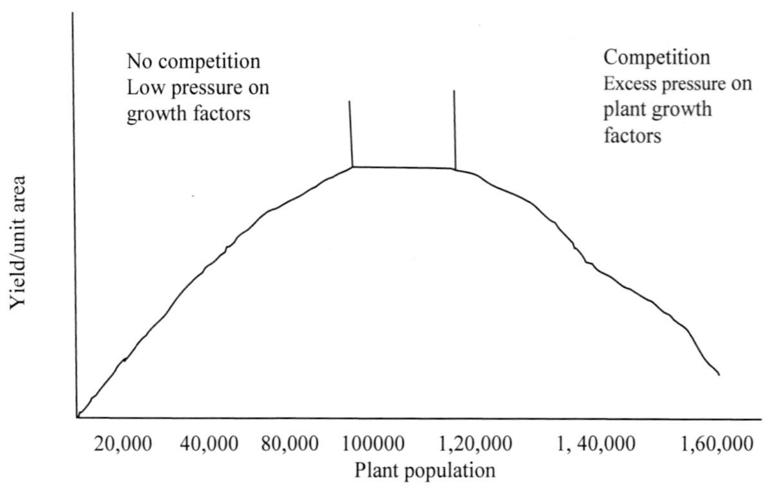

Questions

I. Answer the following

1. What is a seed?
2. What are the qualities of good seed?
3. What are the different types of seed treatments?
4. What are the different methods of sowing the seeds?
5. What are the factors of depth of sowing seeds?
6. What are the problems of seedling establishment?
7. What is the criterion for selection of nursery site?
8. What are the advantages of raising seedlings in the nursery?
9. What are the advantages of square and rectangular planting?

II. Define

1. Dibbling
2. Seed pelleting
3. Plant density
4. Planting geometry
5. Optimum plant density

II. Underline the correct answer

1. Plant propagated by apomictic embryos

   (a) pomogranate (b) mango
   (c) citrus (d) sapota

2. Plant propagated by runner

   (a) citrus (b) straw berry
   (c) garlic (d) gladiolus

3. Propagation through layering

   (a) apple (b) citrus
   (c) cardamom (d) rose

4. Propagation through rooted slips

   (a) zinger (b) banana
   (c) guinea grass (d) sugarcane

5. Propagation through stem cutting

   (a) garlic (b) pineapple
   (c) sugarcane (d) sweet potato

Chapter 8

Crop Nutrition

Criterion for essentiality of plant nutrients

Arnon's (1954) criteria for essentiality of elements for plant growth:

1. The plant must be unable to grow normally or complete its life cycle in the absence of the element,
2. The element is specific and cannot be replaced by another and,
3. The element plays a direct role in metabolism.

Classification of plant nutrients

Carbon, Hydrogen, Oxygen are primary nutrients

Nitrogen, Phosphorus and Potassium are macronutrients

Calcium, Magnesium and Sulphur are secondary nutrients

Iron, Zinc, Manganese, Copper, Boron, Chlorine, Molybdenum, are micronutrients. Nickel and Cobalt are essential for some plants. Hence they are also included under micronutrients.

Sodium, Vanidium, Silicon are called beneficial elements. They play beneficial role in some plants.

Role of nutrients in plant growth

Nitrogen

It is a structural component of chlorophyll, aminoacids and proteins. Plant proteins contain 16 percent N ($6.25 \times \%\ N = \text{protein}$). Plants are dependent on protein for their propagation. Cereal possesses a threshold protein level below which the grains will not be formed. It is supportive for uptake of other nutrients.

Essential for carbohydrate utilization: a certain amount of nitrogen must be present for optimum utilization of carbohydrates produced during photosynthesis. Under deficiency conditions excessive deposition of carbohydrates takes place in vegetative cells with consequent thickening of the cell wall, limited formation of protoplasm, reduced succulence and reduced growth. A growing plant must have continuous supply of free energy input for synthesis of macromolecules from simple precursors and for active transport of ions and other synthesizing materials throughout the plant. Carrier for this energy is adenosine triphosphate (ATP) another indispensable nitrogen containing compound.

It is a component of enzymes. Many enzymes associated with plant growth are all complex proteins containing nitrogen. Growth restrictions results from nitrogen shortages for production of molecules involved in the various enzyme systems.

Hormone molecules which are proteins/ steroids or derivatives of aminoacids serve as coordinators of activity of different plant cells.

It stimulates the root development and activity. Total root mass as well as rooting depth is enhanced by optimal nitrogen availability. Extension of roots facilitates absorption of water and other nutrients required for growth. Classical examples include alleviation of drought effects where deep subsoil moisture is present. Uptake of phosphatic fertilizer is enhanced particularly when NH_4^+-N has been applied. Nitrogen application increases water use efficiency.

Nitrogen directly influences the plant growth. Application of nitrogen increases leaf area. Hence plants are able to trap more light by increasing both the rate at which the cells multiply and thence their number.

In cereals nitrogen increases the number of tillers/side shoots. Excess of nitrogen causes self shading. The lower leaves do not receive sufficient light. They turn pale green or yellow. They do not photosynthesize efficiently but continue to respire so that there is net loss of assimilates to the plants. Nitrogen responsive varieties should have oblique or acute leaf angle.

Leaf area duration (LAD): Maintenance of leaf area produced for longer period to take best advantage of the season. Nitrogen along with phosphorus plays dominant role in maintenance of leaf area.

- Nitrogen increases the total dry matter and improves the yield parameters and yield.

  Nitrogen improves the quality of crops:
- Nitrogen fertilization improves the protein quality and bread making quality of wheat. Milling quality of rice.
- Excess nitrogen application in potato increases total drymatter but reduces the tuber yield.
- Excess nitrogen reduces sugar production in sugarbeet. Oil content in oilseeds is reduced, but improves the protein content.
- In crops where the green leaves are the economic part of the plant improves the crop productivity. *e.g.*, cabbage, fodder, leafy vegetables. Excess of nitrogen application to green vegetables may result in NO_3 accumulation in the leaves.

- Excess of nitrogen results in formation of weak fibres in cotton and jute. The excess nitrogen is utilized in synthesis of proteins in the presence of excess potassium.
- Quality of fruits is impaired. Quality of tobacco deteorates due to more nicotine content under excess nitrogen application.

Disease and pest incidence and other ill effects of excess nitrogen

- Plants become excessively succulent and become susceptible to diseases *e.g.*, blast
- If phosphorus, potassium and sulphur are deficient in soil, excess of nitrogen may delay maturity.
- In cereals lodging may occur- due to increase in height of straw/weak stem due to inadequate accumulation of celluloses.
- In legumes excessive nitrogen reduces nodulation.
- Excess nitrogen induces copper and zinc deficiency.

Deficiency occurs due to actual low level of nitrogen in the soil and imbalance of nutrients. Under actual nitrogen deficient conditions plants will appear stunted, pale yellow. The yellowing appears first on the lower leaves. In severe cases leaves will turn brown and die. When nitrogen is deficient in plants the nitrogen compounds in the older leaves undergo lysis. The protein nitrogen is converted into soluble form and translocate to active meristematic region. Nitrogen is the first element to become deficient in arid and semiarid regions of low rainfall and rapid decomposition of organic matter, also in sandy soils in heavy rainfall areas. It is mobile element in plants.

Phosphorus

Phosphorus is the component of adenosine di and tri phosphate (ATP and ADP). ATP is involved in storage and transfer of energy. Transfer of energy from ATP to energy requiring molecule is called phosphorilation. Regeneration of ATP from ADP requires phosphorus at the sites of energy production.

Phosphorus is the structural component of nucleic acid, coenzymes, nucleotides, phosphoproteins, phospholipids and sugarphosphates. Adequate supply of phosphorus during the early stage of crop growth is required for laying down primordia for reproductive parts. Phosphorus is found in large quantities in growing parts, fruits and seeds. Phytin is the principal storage form of phosphorus in seeds. Phospholipid is another phosphorus containing compound forms the structure of protoplasm.

Phosphorus is associated with root growth, root proliferation. Adequate supply of phosphorus induces tolerance to root rot disease, winter damage or cold injury, greater strength to cereal straw, improves the quality of fruits, forages, vegetables and grain crops, early maturity of crops.

Deficiency of phosphorus results in slow emergence and growth, off colour green foliage with purple venation especially on the underside of the older leaves. Petioles have a purple cast, root development is poor, decreased leaf growth, narrow leaves with tips green-violet or pinkish or greenish blue due to anthocyanin accumulation under severe deficiency, there will be drying of leaf tips and margins, decreased tillering, decreased photosynthetic activity and decreased translocation of photosynthates, excess accumulation of sugars leads to bronzing and purpling of leaves in maize. Biological nitrogen fixation is limited by phosphorus deficiency. Phosphorus deficient plants look spindly and stunted.

Excess of Phosphorus: induce copper and zinc deficiency. The interaction of phosphorus with copper, zinc and iron are antagonistic.

Potassium

Involved in activating the enzymes (as many as 80 enzymes), responsible for carbon dioxide reduction. facilitates translocation of photosynthates, nitrogenous compounds from vegetative parts to the grain thereby increases protein content of grain. Increase in nitrogen fixation by Rhizobia in legumes with adequate supply of potassium is ascribed to greater translocation of photosynthates from source to the nodules, besides it activates the nitrogenase enzyme which is responsible for reduction of N_2 to NH_4 in the cells of the Rhizobium. Adequate supply of potassium promotes formation of stronger supporting tissue which is the cause for reduced lodging in cereals. Improves the storability of fruits, imparts resistance to fungal diseases.

Deficiency of potassium reduces crop growth and yield.

Calcium

Calcium imparts structure and permeability to cell membrane. Calcium enhances the uptake of NO_3–N and interrelated with NO_3 metabolism. It regulates cation uptake, cell elongation and cell division. Structural component of chromosome, Calcium is required for rhizobium infection in legumes neutralizes organic acids *viz.* citric, malic and oxalic acids.

Deficiency of calcium causes general breakdown of membrane structure, failure of terminal buds and apical tips of roots to develop resulting in cessation of growth, failure of leaves to unfold, blossom end rot of tomato, bitter pit of apple.

It is immobile in plant. Very little translocation of calcium observed in the phloem. Hence there is poor supply of calcium to fruits and storage organs. Downward translocation of calcium is also limited. Hence roots fail to enter in soils deficit in calcium.

Magnesium

Magnesium is the structural component of chlorophyll and ribosomes. Chlorophyll alone accounts for 15–20% of the total magnesiumin plants. Ribosomes are involoved in the synthesis of protein. Hence magnesium deficiency reduces protein synthesis and increases non protein nitrogen in the plants. It is involved in the activity of every phosphorilating enzyme in carbohydrate metabolism, responsible for energy transfer and metabolic process in the plants. Increases oil content in oil seeds.

Deficiency of magnesium leads to interveinal chlorosis in leaves where in only the veins remain green. In cotton lower leaves turn reddish-purple cast due to accumulation of anthocyanin pigments which gradually turn brown and finally necrotic. Premature leaf shedding. Maize, potato and cucurbits are the indicator plants. Cattle consuming magnesium deficient forage crops may suffer from Hypomagnesemia or grass tetany. Application of high rates of NH_4^+–N or potassic fertilizer induce grass tetany.

If the magnesium concentration in the mature leaves is 0.20 to 0.25 percent the plants do not show magnesium deficiency.

Sulphur

It is a constituent cysteine and methionine, vitamins, lipoic acid, acetyl co enzyme A, involved in synthesis of amino acids and fatty acids. Important in oilseed, pulses, root and starch crops.

Deficiency of sulphur causes chlorosis of younger leaves, stunted growth, stem girth reduces, plant appear spindly.

Zinc

It plays significant role in assimilation of NO_3-N and synthesis of tryptophane. Involved in biosynthesis of IAA which is responsible for flowering and fruiting, it is an activator of several enzymes (triphosphate dehydrogenase and tryptophan synthetase), involved in photosynthesis and nitrogen metabolism.

Deficiency of zinc results in poor growth of the terminal bud. As a consequence there will be decrease in stem length (stunted growth) resulting in resetting and whorling or rosette of leaves leading to bushy appearance of the plants. Zinc deficiency disorder in rice is called *khaira* (which is common when zinc availability in soil is < 0.7 ppm).white bud in maize and malforming of fruits, There will be reduction in water uptake. Interveinal chlorosis often with necrosis and pigmentation and reduced leaf size occurs in broad leaved crops known as little leaf and malformation of leaves.

Iron

It plays catalytic role in the activities of several enzymes, chlorophyll synthesis pigmentation and nitrogen fixation, photosynthesis and respiration.

Deficiency causes interveinal chlorosis in younger leaves under deficient situation, leaves may turn white.

Boron

Required for development and growth of new tissues particularly vascular elements, carbohydrate and nucleic acid metabolism, translocation of photosynthates as sugarborate complex.

Deficiency of boron is associated with abnormal or retarded apical growth or resetting, die-back, discoloration or failure to grow or elongate and stimulates lateral growth of plants. Deficiency is observed in the region of high metabolic activity. Young leaves become thicker, wrinkled darkish blue green colour. Irregular chlorosis between midrib and the interveins, leaves and stems become brittle. Thickening of petioles or stems, corky, cracked or cross hatched or water soaked dead areas. In fruits, tubers or tuberous roots the fleshy parts may show brown flecks, necrosis, cracks or dry rot, may be water soaked or may show discolouration in the vascular system. Boron deficiency disorder is called by different names in different crops

Sugarbeet - heart rot

Tobacco - top sickness

Apple- cark disease

Cauliflower-brown rot/ hallow stem

Sunflower- malformed capitulum, with unfilled regions

Groundnut- hollow heart

Potato-brown discolouration in the vascular region

Copper

It is a part of enzymes involved in cellular oxidation- reduction, electron transport during photosynthesis and carbon assimilation and involved in several metabolic process.

Deficiency of copper leads to chlorosis, dieback of terminal bud, shortened internodes, splitting of young fruits observed in orange, cabbage, lettuce; poor or absence of heading; withering and distortion of apices of younger leaves under extreme deficiency of copper in cereals and legumes in recently reclaimed soils which is called reclamation disease.

Manganese

Manganese is required for chlorophyll formation and development of chloroplast. Essential for respiration, nitrogen metabolism and photosynthesis activator of variety of enzymes concerned with hydrolysis.

Deficiency of manganese leads to dark green bands along the midrib and main veins with lighter green areas between the bands. In severe case dark green areas tends to dull green and light green while light green area becomes white. The twig bearing such foliage shows dieback symptom. Chlorosis necrotic spots in interveinal areas in cereal leaves, mottling in dicot leaves is observed. Copper deficiency disorder in crops is called by different names.

Grey specks -Oats

Speckled yellows –Sugar beet

Marsh spot –Peas

Pahala blight- Sugarcane

Molybdenum

It is the constituent of nitrate reductase, which is responsible for NO_3-N metabolism in plants and protein synthesis. Hence plants appear dark green due to accumulation of NO_3-N in the leaves. Involved in carbohydrate metabolism, associated with symbiotic nitrogen fixation in nodules,

Deficiency of molybdenum causes marginal scorching, rolling or cupping of leaves, chlorotic mottling between the veins of older and middle leaves all over the surface, necrotic under severe deficiency, inhibition of flower formation, nodule development and nitrogen fixation, Molybdenum deficiency disorder is called whiptail in cauliflower, yellow spot in citrus, downward cupping in radish and scald in beans.

Chlorine

It is involved in oxygen evolution during photosynthesis and cyclic photo-phosphorylation.

Deficiency of chlorine leads to wilted appearance and stuffy roots with profusely branched laterals, chlorotic mottling, bronzing and necrosis of tissues.

Questions

I. Answer the following

1. What is the criterion for essentiality of plant nutrient?
2. Classify the plant nutrients.
3. What is the role of nitrogen in plants?
4. What is the significance of nitrogen on the quality of crops?
5. How does plants responds to excess of nitrogen application?
6. What is the role of phosphorus in plants?
7. What is the significance of potassium in plant growth?

8. What is the role of calcium, magnesium and sulphur in plants?
9. State the role of micronutrients in plants?
10. What are the deficiency disorders of different micronutrients in plants?

II. Underline the correct answer

1. Essential element involved in increasing the leaf area and dry matter production in plants
   (a) phosphorus (b) nitrogen
   (c) potassium (d) calcium
2. Beneficial element in plants
   (a) Nitrogen (b) potassium
   (c) Silicon (d) calcium
3. Excess application of __________ promotes disease incidence in plants
   (a) nitrogen (b) phosphorus
   (c) potassium (d) magnesium
4. An element which is a component of ATP
   (a) nitrogen (b) phosphorus
   (c)potassium (d) calcium
5. An element involved in activity of large number of enzymes.
   (a) phosphorus (b) potassium
   (c) nitrogen (d) sulphur
6. An element whose downward translocation is limited.
   (a) magnesium (b) nitrogen
   (c) calcium (d) potassium
7. Interveinal chlorosis is the deficiency disorder of
   (a) Calcium (b) nitrogen
   (c) sulphur (d) magnesium
8. Constituent of acetyl Co enzyme A is
   (a) Calcium (b) sulphur
   (c) magnesium (d) Phosphorus
9. *Khaira* is the deficiency disorder of
   (a) iron (b) manganese
   (c) zinc (d) copper
10. An immobile element in the plant
    (a) iron (b) potassium
    (c) nitrogen (d) phosphorus
11. Dieback in citrus is due to deficiency of
    (a) iron (b) copper
    (c) molybdenum (d) calcium

12. An element involved in translocation of photosynthates in plants is
    (a) iron (b) manganese
    (c) copper (d) boron
13. Pahalablight is the deficiency disorder of
    (a) manganese (b) iron
    (c) magnesium (d) copper
14. An element which is constituent of nitrogenase and nitrate reductase
    (a) calcium (b) magnesium
    (c) molybdenum (d) boron
15. An element involved in oxygen evalution during photosynthesis
    (a) chlorine (b) manganese
    (c) zinc (d) copper

Chapter 9

Manures and Fertilizers

Manures are the plant and animal wastes that are used as source of plant nutrients. They release nutrients upon their decomposition.

Bulky organic manures contain small percentage of nutrients and they are applied in large quantities. *e.g*,. farmyard manure, compost, green manures.

Farmyard manure: is a mixture of dung and urine of the farm animals along with litter, left over material from roughages or fodder fed to the cattle. It has 0.5% N, 0.2% P_2O_5 and 0.5% K_2O.

Compost: Decomposed mass of organic wastes. If the wastes are from farm, it is referred as farm compost. It has 0.5% N, 0.15% P_2O_5 and 0.5% K_2O. If the refuses *viz.* street sweepings, night soil, dust bin wastes are used it is referred as town compost. It has 1.4% N, 1.0% P_2O_5 and 1.4% K_2O.

I. Composition of animal excreta

S. No.		N	P_2O_5	K_2O
1.	Cattle dung	0.15	0.10	0.05
2.	Buffalo dung	0.15	0.10	0.05
3.	Cattle dung and urine mixed	0.60	0.15	0.45
4.	Sheep/Goat dung	0.65	0.50	0.03
5	Sheep dung and urine mixed	0.95	0.35	1.0
6.	Pigs dung	0.60	0.50	0.03
7.	Poultry excreta	0.80	0.60	0.30
8.	Human excreta	1.60	1.20	0.55

II. Composition of bulky organic manures

1.	Farm yard manure (FYM)	0.5-1.5	0.4-0.8	0.5-1.9
2.	Compost	1.0-2.0	1.0	1.5
3.	Green manures (on average)	0.5-0.7	0.1-0.2	0.6-0.8
4.	Night soil	1.2-1.3	0.8-1.0	0.4-0.5

Green manures

Green manure is defined as fresh organic matter/undecomposed plant material/ real organic matter, natural plant food added to the soil for the purpose of supplying plant nutrients.

Types of green manuring

Green manuring: When the green manure crop is grown in the field where it is to be incorporated is known as green manuring or *in situ* green manuring. The seeds may be sown in the field either before sowing the main crop or along with the main crop and incorporated into the soil at flower initiation or at 45–60 days of sowing.

1. Green manure plants may be leguminous or nonleguminous
   (a) Leguminous green manures
   Sunnhemp (*Crotalaria juncea*)
   Dhaincha (*Sesbania aculeata*)
   Guar (*Cyamopsis tetragonoloba*)
   Cowpea
   Green gram
   (b) NonLegumes *e.g.*, oats, barley, maize and sorghum
2. Green leaf manuring: The practice of collecting green leaves from all the available plant sources and incorporating into the soil is called green leaf manuring (*ex situ* manuring)
   e.g., Glyricidia (*Glyricidia maculata*)
   Sesbania (*Sesbania speciosa*)
   Karanj (*Pongamia Pinnata*)
   Neem (*Azadirachta indica*)

Nutrient content of green manures on dry weight basis (%)

Green manure	N	P_2O_5	K_2O
Sesbania/Dhaincha	3.5	0.6	1.2
Sunnhemp	2.3	0.5	1.8
Cowpea	2.3	0.5	2.1
Guar	3.0	0.4	1.6
Pongamia	3.3	0.4	2.4
Glyricidia	2.9	0.5	2.8
Calatropis gigantea	0.42	0.12	0.37
Wild indigo	3.2	0.3	1.3
Sesbania rostrata	3.1	0.3	1.46

Advantages of Greenmanuring

- Adds organic matter
- Conserves nutrients
- Mobilization of nutrients from deeper layers
- Availability of phosphorous increases

- Leguminous green manures improve the availability of nitrogen from humus
- *In situ* green manuring smothers weeds
- Green manure crops control root knot nematodes
- They serve has catch/ forage/shade/ fibre crop

Limitation of *in situ* green manure

- Loss of one season
- Additional cost of cultivation: (cost of seeds, sowing, irrigation)

Characteristics of *in situ* green manure plants

The green manure plant should be multipurpose, fast growing, high nutrient accumulating, tolerant to flood, shade, drought pests and diseases. They should have weed suppressing ability, wide adaptability, high water use efficiency, early onset of biological nitrogen fixation, photoperiod insensitive, soft, succulent foliage, high seed multiplication, longer seed viability and response to microbial inoculation.

Concentrated organic manures

These are organic manures which contain higher concentration of nutrients

1. Oilcakes-these are the byproducts of oil industry.

   They may be edible or non edible.

   Edible oil cakes-groundnut cake, sunflower cake, niger cake, safflower cake, sesamum cake.

   Non edible oilcakes–These are not fit for feeding livestock *e.g.*, castor cake, neem cake, mahua cake.

   The nutrient content in oil cakes varies from 2.5–7.9% N, 0.8–2.9% P_2O_5 and 1.2–2.2% K_2O. The nutrients present in oilcakes are made available upon microbial degradation in soil within 7–10 days of application.
2. Fish meal/Fish manure–Excess of fish caught during the season if unsold are dried, powdered which is known as fish manure. On an average the fish manure contains 4–10% N, 3–9% P_2O_5 and 0.3–1.5% K_2O.
3. Bone meal- Bones of the slaughtered animals are collected dried and grounded into powder is known as raw bone meal. It contains 3-4% N and 20–25% P_2O_5. If the bones are steamed under pressure for removing fat which is used in the preparation of soap and candles and nitrogen extracted which is used in the manufacture of glue. The residual matter is grounded into powder is called steamed bone meal. It contains 1–3% N and 22–24% P_2O_5.

Nutrient content of Non edible oil cakes

	N	P_2O_5	K_2O
Castor cake	4.3	1.8	1.3
Neem cake	5.2	1.0	1.4
Cotton (undecorticated)	1.8	1.6	3.9
Karanj (Pongamia) cake	4.2	0.9	1.20
Mahua(Ippi) cake	2.5	0.8	1.8
Safflower (undecorticated)	4.9	1.4	1.2

Nutrient content of Edible oil cakes

Groundnut cake	7.3	1.5	1.3
Coconut cake	3.0	1.9	1.8
Linseed cake	4.9	1.4	1.3
Niger cake	4.7	1.8	1.3
Rapeseed cake	5.2	1.8	1.3
Sesamum cake	6.2	2.0	1.2
Cotton seed cake (decorticated)	6.4	2.9	2.2
Safflower seed cake (decorticated)	7.9	2.2	1.9
Blood meal	10–12	1–2	1.0
Meat meal	10–15	2.5	0.5
Fish meal	410	3–9	0.3–1.5
Horn & Hoof	13		
Raw bone meal	3–4	20–25	-
Steamed bone meal	1.2	25–30	-
Bird Guana	7-8	11–14	2–3

Fertilizers

These are industrially manufactured chemicals containing plant nutrients. Nutrients are immediately released upon application to soil.

Classification of fertilizers

1. Based on the number of primary nutrients present
   (a) Straight fertilizers: are those which supply only one primary plant nutrient *e.g.*, urea, single superphosphate, muriate of potash.

   Straight fertilizers are further grouped into

   (i) Nitrogenous fertilizers *e.g.*, urea

   (ii) Phosphatic fertilizers *e.g.*, single superphosphate

   (iii) Potassic fertilizers *e.g.*, muriate of potash

   (b) Complex fertilizers: Containing two or more primary plant nutrients which are in chemical combinations. These fertilizers are in granular form *e.g.*, Diammonium phosphate (DAP), nitrophosphates, ammonium phosphates. Complex fertilizer containing two of more primary plant nutrients but they are not in chemical combination. *e.g.*, 17-17-17, 19-19-19, 20-20-0.

2. Fertilizers are classified into low analysis fertilizers and high analysis fertilizers based on the concentration of primary nutrient (NPK).
   (a) Low analysis fertilizers: They contain < 25% primary nutrient. e. g. Single super phosphate 16% P_2O_5; Sodium nitrate 16% N.
   (b) High analysis fertilizers: The total nutrient content of primary nutrients is >25% *e.g.*, Urea 46% N, Muriate of potash 60% K_2O, Ammonium phosphate: 20% N + 20% P_2O_5, DAP 18% N+ 46% P_2O_5
3. Fertilizers can also be classified based on physical form
   (a) Solid fertilizers
   (b) Dust: single superphosphate
   (c) Crystals: Ammonium sulphate
   (d) Prills: Urea, DAP,
   (e) Briquettes: Urea briquettes
   (f) Clear liquid suspensions
   (a) Nitrogenous fertilizers are classified into four groups based on the chemical form of nitrogen in the fertilizer .
      i. Nitrate fertilizers: These contain nitrogen in the form of nitrate (NO_3)
         e.g., Sodium nitrate (16% N)
         Calcium nitrate (15.5% N & 19.5% Ca)
         Potassium nitrate (13.8% N & 36.5% K_2O)
      ii. Ammonical fertilizers: These contain nitrogen in the form of ammonium (NH_4)
         e.g., Ammonium sulphate (20% N and 24.2% S)
         Ammonium Chloride (24 to 26% N)
         Anhydrous ammonia (NH_3) (82% N)
      iii. Nitrate and ammonical fertilizers: These contain nitrogen in the form of both nitrate and ammonium.
         e.g., Ammonium nitrate (33–34% N)
         Calcium ammonium nitrate- (20% N)
      iv. Amide fertilizers contain nitrogen in the form of amide. These are also known as organic fertilizers since they contain carbon atoms.
         e.g., Urea ($CO(NH_2)_2$) (46% N)
         Calcium Cyanamide ($CaCN_2$) (21% N)

(b) Phosphatic fertilizers: Classified into three groups based on the solubility and availability of phosphorus present in the fertilizer.

i. Phosphatic fertilizers containing water soluble phosphoric acid.

e.g., Single super phosphate (16% P_2O_5)

Monoammonium phosphate (11% N, 48% P_2O_5)

Diammonium phosphate (18% N, 46% P_2O_5)

ii. Phosphatic fertilizers containing citric acid soluble phosphoric acid or dicalcium phosphate.

e.g., Basic slag (14–18% P_2O_5)

Dicalcium phosphate (34–39% P_2O_5)

iii. Phosphatic fertilizers containing phosphoric acid not soluble in water and citric acid.

e.g., Rock phosphate - 20–40% P_2O_5

Raw bone meal 20–25% P_2O_5

Steamed bone meal 22% P_2O_5

These are suitable for plantation crops.

(c) Potassic fertilizers

Potassium in potassic fertilizers exists in combination with chloride, sulphate or nitrate.

Potassium chloride (KCl) contains 60% K_2O

Potassium sulphate (K_2SO_4) contains 50% K_2O

Potassium nitrate (KNO_3) contains 13% N and 44% K_2O

Sources of secondary nutrients

Sulphur Single super phosphate (13.9% S)

Ammonium sulphate (24% S)

Potassium sulphate (18% S)

Gypsum (18.6% S)

Pyrites (16% S)

Elemental sulphur (85 to 100% S)

Copper sulphate (12% S)

Calcium Single superphosphate (18–21%Ca)

Triple super phosphate (12–14% Ca)

Calcium nitrate (19% Ca)

Gypsum (23% Ca)

Liming materials–Calcite, dolamite, hydrated lime, blast furnace slag

Magnesium-	Magnesium nitrate (16% Mg) Magnesium chloride (8–9% Mg) Magnesium sulphate (9.8% Mg) Magnesia (MgO) (55% Mg) Dolamite lime stone.

Sources of micronutrients

Zinc	Zinc sulphate (23% Zn) Zinc oxide (78% Zn) Zinc Carbonate (52% Zn) Zinc sulphide (67% Zn) Zinc Chelates: Na_2 Zn EDTA (14% Zn), Na Zn NTA(13% Zn), Na Zn HEDTA (9% Zn)
Boron	Borax (11% B) Boric acid (17% B)
Iron	Iron chelates–Fe–EDTA, Fe–HEEDTA, Fe- DTPA, Fe, EDDHA
Manganese	Manganese sulphate (26–28%) Manganese Chloride (17%) Manganese Oxide (41– 68%) Chelates of Manganese–Mn-EDTA (5-12% Mn)
Copper	Copper sulphate (25% Cu) Copper oxide (75% Cu) Chelates of copper -Na Cu- EDTA (13% Cu), Na-Cu HEDTA (9% Cu)
Molybdenum	Ammonium Molybdate (54% Mo) Sodium Molybdate (38% Mo)

Biofertilizers

Biofertilizers are the materials containing microbial cultures which upon inoculation to seeds/seedlings or application to the soil fix atmospheric nitrogen either symbiotically or asymbiotically, they solubilize insoluble nutrients present in the soil or mobilize physically inaccessible plant nutrients to the plants. These are complementary or supplementary sources of plant nutrients.

Types of biofertilizers

Nitrogen fixers

Symbiotic nitrogen fixers: Bacteria fixes atmosphonic nitrogen in association with leguminous host plants by producing nodules in roots or stem or branches.

The host plant will supply photosynthates to nitrogen fixing bacteria while bacteria will provide fixed nitrogen to plants *e.g.*, Rhizobia.

Certain group of leguminous plants shows specificity or preference to single kind of Rhizobium. The bacteria that nodulate plants belonging to these groups are designated as 'Species'. All plants nodulated by a rhizobium species constitute a 'cross' inoculation group.

Cross inoculation groups

Rhizobium species	**Cross inoculation group**	**Legumes**
R. *trifoli*	Clover group	Trifolium
R. *meliloti*	Alfalfa group	Melilotus, Medicago
R. *phaseoli*	Bean group	Pheseolus
R. *lupine*	Lupine group	Lupinus, ornithopus
R. *leguminosarum*	Pea group	Pisum, Vicia, Lens
R. *japonicum*	Soybean group	Glycine
Rhizobium sp (miscellany)	Cowpea	Vigna and Arachis

Nitrogen fixing potential of legumes:

Crop	N fixed (kg per ha)
Pigeonpea	68–200
Soyabean	49–130
Groundnut	112–152
Chickpea	26–63
Cowpea	53–85
Greengram	50–55
Alfalfa	100–300

Asymbiotic N fixers: fix atmospheric nitrogen in the rhizosphere of nonlegume plants by diazotrophs is called associative nitrogen fixation or rhizocoenoses.

e.g., *Azotobacter chroococcum*

Azospirillum lipoferum

Azospirillum brasilense

The extent of fixed nitrogen ranges from 25 to 35 kg per ha

Phosphorus solubilising microorganism

Microbial agent that converts the fixed forms (insoluble phosphorus) of phosphorus (ferric phosphate, aluminium phosphate or tricalcium phosphate) into available ionic forms ($H_2PO_4^{-1}$, HPO_4^{-2}) are called mineral phosphate solubilizers.

Bacteria: *Agrobacterium radiobacter, Bacillus polymyxa,, Bacillus megatherium.*

Fungi: *Aspergillus awamorii, Aspergillus flavous, Aspergillus niger*

Questions

I. Answer the following

1. What are the bulky organic manures?
2. Differentiate farmyard manure and compost
3. Classify the green manures.
4. What are the advantages of green manuring?
5. What are the characteristics of *in situ* green manuring plants?
6. What are concentrated organic manures?
7. What are fertilizers?
8. Classify the nitrogenous fertilizers
9. Classify phosphatic fertilizers.
10. What are biofertilizers?
11. What are cross inoculation groups of rhizobia?
12. What are phosphate solubilizing microorganisms?

II. Underline the correct answer

1. Cowpea cross inoculation group nodulates
   (a) *Glycene max* (b) *Arachis hypogaea*
   (c) *Trifolium* sp. (d) *Melilotus* sp.
2. *Medicago sativa* is nodulated by
   (a) *Rhizobium trifoli* (b) *Rhizobium meliloti*
   (c) *Rhizobium Phaseoli* (d) *Rhizobium japonicum*
3. *Rhizobium leguminosarum* nodulates
   (a) Soyabean (b) Peas
   (c) groundnut (d) Cowpea
4. Zinc content of zinc sulphate is
   (a) 78% (b) 23% (c) 50% (d) 67%
5. Boron content of borax is
   (a) 17% (b) 20% (c) 11% (d) 24%
6. Sulphur content of gypsum is
   (a) 18.6% (b) 24% (c) 13.9% (d) 12%
7. The P_2O_5 content of single super phosphate is
   (a) 21% (b) 16% (c) 48% (d) 24%

8. Urea contains
   (a) Ammonical form of nitrogen
   (b) Amide form of nitrogen
   (c) Nitrate form nitrogen
   (d) Ammonical and nitrate form of nitrogen
9. An high analysis fertilizer
   (a) single superphosphate (b) muriate of potash
   (c) sodium nitrate (d) calcium cyanamide
10. Bone meal contains
    (a) water soluble phosphorus
    (b) citrate soluble phosphorus
    (c) water and citrate insoluble phosphorus
    (d) water and citrate soluble phosphorus
11. The nitrogen content of neem cake is
    (a) 5.2% (b) 7% (c) 3% (d) 1.5% K_2O
12. The nutrient value of sunnhemp green manure on dry weight basis
    (a) 2.3%N, 0.5% P_2O_5 and 1.8%
    (b) 3.3%N, 0.4% P_2O_5 and 2.4% K_2O
    (c) 0.42%N, 0.12% P_2O_5 and 0.37% K_2O
    (d) 0.50%N, 0.2% P_2O_5 and 0.5% K_2O
13. An example of *in situ* green manure
    (a) *Glyricidia* (b) *Pongamia*
    (c) *Azadiricta indica* (d) *Crotalaria juncea*
14. Nutrient value of farmyard manure
    (a) 0.50%N, 0.20% P_2O_5 and 0.5% K_2O
    (b) 1.4%N, 1.0% P_2O_5 and 2.4% K_2O
    (c) 0.15%N, 0.10% P_2O_5 and 0.05% K_2O
    (d) 0.8%N 0.60% P_2O and 0.3% K_2O
15. Nutrient value of fish meal
    (a) 4–10%N, 3–9% P_2O_5 and 0.3–1.5% K_2O
    (b) 3–4%N, 20–25% P_2O_5 and 1.5% K_2O
    (c) 0.50%N, 0.20% P_2O_5 and 0.50% K_2O
    (d) 4.30%N, 1.80% P_2O_5 and 1.30% K_2O

Chapter 10

Fertilizer Management in Crop Production

Knowledge of fertilizer dose, selection of fertilizer sources, time and method of application is essential for achieving efficient use of fertilizers and profitable yield.

A. Fertilizer Dose

Fertilizer dose for crops is determined based on the field experimentation. It varies with the crops and varieties, initial fertility of the soil, soil properties, cropping system, climate, season, rainfed or irrigated crop, expected yield *etc.*

Fertilizer dose for rainfedcrops (Kg/ha)

Fingermillet	50–35–25
Redgram	25–50–25
Groundnut	25–50–25

Fertilizer dose for irrigated Crops (Kg/ha)

Fingermillet	100–50–50
Paddy	100–50 - 50
Maize	150–75–25
Groundnut	25–75–25

At optimum fertilizer dose, the crop will produce maximum economic yield, whereas at maximum crop yield the fertilizer dose may be highest which may not be economical.

B. Selection of nutrient sources/ fertilizers

Selection of the fertilizers depends on the type of crops, effect of fertilizers on the crop quality, type of soil and climate.

1. Ammonical form of nitrogenous fertilizer is appropriate for submerged paddy.
2. In soils rich in calcium carbonate application of ammonical form of nitrogenous fertilizer results in loss of nitrogen in the form of ammonia. Hence nitrate form of nitrogenous fertilizers is appropriate.

3. Nitrogenous fertilizers containing both ammonical and nitrate form of nitrogen are ideal for most of the crops
4. Phosphatic fertilizers with water soluble phosphorus are suitable for neutral and basic soils, while citrate and citrate insoluble phosphatic fertilizers are appropriate for acidic soils and long duration crops.
5. Combination of water and citrate soluble phosphatic fertilizers are suitable for medium and long duration crops, while water soluble phosphatic fertilizer is ideal source for short duration crops.
6. In crops *viz.* tobacco and potato, sulphate of potash is appropriate. Use of muriate of potash spoils the quality of crops due to the presence of chlorine.

C. Time of fertilizer application

Basal dose

Plants do not absorb the entire dose of nutrients at a single time. The nutrient supply should match with the crop demand. Further if large quantities of nutrients are supplied at a single time the nutrients are prone to losses. Nitrogen is highly dynamic subject to transformation into gaseous forms *viz.* N_2O, NO, NH_3 and elemental nitrogen, which will be lost to atmosphere. To minimize the losses, fertilizers are applied in split doses. Part of the recommended fertilizer is applied prior to or at sowing or at planting the crop is called basal dose. Generally 50 percent of the recommended nitrogen is applied at sowing. In sugarcane where recommended dose is large and initial crop growth is slow 10 percent of the recommended nitrogen is applied at planting.

Nutrient sources *viz.* bulky organic manures, green manures, phosphatic fertilizers containing di and tricalcium phosphate (*e.g.*, Rock phosphate and bone meal), elemental sulphur, pyrites should be applied at least three weeks before sowing the crop to facilitate decomposition and release of nutrients by microbial action.

Entire dose of phosphatic and potassic fertilizers, sulphur, calcium and magnesium sources are applied as basal dose. Micronutrient sources are used in smaller doses; hence they are diluted with loamy soil or well decomposed compost and applied to soil at sowing. Seed treatment or root dipping in transplanted crops in solution or slurry of micronutrients may be followed. Biofertilizers are applied at sowing by mixing with well decomposed compost closer to the seeds or applied as seed treatment in leguminous crops.

Top dressing

Fertilizers applied after sowing or planting/transplanting the crop is termed as top dressing. Generally nitrogenous fertilizers are top dressed. Top dressing may be done once, twice or three times depending upon the duration of the crops. In rice two top dressings of nitrogenous fertilizers is followed while, in sugarcane three top dressings are recommenced at an interval of three to four weeks from the date of planting to match with the crop demand.

D. Methods of Fertilizer Application

Solid form fertilizers

Broadcasting: Spreading of manures/fertilizers uniformly all over the field is called broadcasting. Fertilizers may be broadcasted and left on the surface as in perennial forage crops or under no- till system. Broadcasted fertilizers are incorporated by ploughing/ harrowing /cultivating.

Advantages

Large quantity of fertilizer may be applied without injuring the crop, nutrients are uniformly incorporated throughout the plough layer, less labour requirement.

Disadvantages

Nutrients are easily accessible to weeds, water soluble phosphatic fertilizers are readily converted into insoluble form in acidic and calcarious soils, fixation of potassium and ammonical form of nitrogen will be greater in soils containing 2:1 type of clay mineral. Not suitable for wider spaced crops.

Broadcasting may be adopted for forage and pasteur crops, for crops whose roots are uniformly distributed in surface layer (rice, onion), for application of citrate soluble phosphatic fertilizers *e.g.*, bone meal, basic slag and rock phosphate.

Placement of fertilizers: Placed at particular site relative to seed or seedling or crop row or plant.

Plough-sole placement: Adopted in dry soil where there is moisture in the plough sole layer and in problematic soils, or in soils where there is the problem of ammonia, phosphorus and potassium fixation. Manures and fertilizers are placed in the plough-sole after opening the furrow which is covered by the subsequent furrow slice.

Deep placement: Ammonical fertilizers are placed in the reduced layer in puddled soil and water soluble phosphatic fertilizers in black soils to increase efficacy.

Localized placement

Contact placement or combine drilling: Fertilizers are drilled along with the seed using suitable equipment such as seed sum fertilizer drill. Fertilizers are placed in narrow bands by the side of seed rows at the desired depth. In majority of the field crops, fertilizers are placed 5-8 cm away from the crop rows on either side or at a depth 2–5 cm below the seed row.

Placement is particularly useful in maintaining the solubility of the phosphates. When soluble phosphates are broadcasted and thoroughly mixed with the soil, they are rendered insoluble. If they are placed in bands, the peripheral layer of phosphates is fixed by the soil and the core of phosphates inside the bands remains unaffected. The soil surface in contact with the fertilizer is limited and it fixes a certain quantity. This layer of phosphorus satisfied soil protects the rest of the water soluble phosphorus.

Band placement: Adopted in row sown crops. It is also called row placement. Bands may be continuous in crops where intra row spacing is lesser (sorghum) but discontinuous in crops where row spacing is more (cotton). Ring or hill placement is adopted in widely spaced crops (cotton, chillies, sugarcane). This is also called side dressing.

Pocket placement: fertilizers are placed deeper into the pocket (dibbled) and seeds are sown 5 cm above the fertilizer or manurial mixture in the same pocket. Paper packets of urea, large granular urea and urea super granules are placed by this method in paddy.

Pellet application: Urea or ammonium sulphate mixed with the clay soil in 1:10 and made into a dough. The Pellets or the balls of convenient size are prepared and placed at the centre of the four hills of rectangular or square planted paddy to a depth of 5-7 cm. Pellet placement minimizes denitrification loss of nitrogen.

Liquid form application

Starter solutions: Solutions of fertilizers prepared in low concentrations which are used for soaking seeds, dipping roots or plants after uprooting or sprayed on the seedling to strengthen the seeds or seedling for early rooting establishment and growth. Paddy seeds presoaked in 40 percent potassium dihydrogen phosphate (KH_2PO_4) solution for 12 hours. Paddy or finger millet seedlings are dipped in phosphate slurry, zinc sulphate solution. Solutions of N: P_2O_5:K_2O at 1:2:1 or 1:1:2 applied to the young vegetable plants at the time of transplanting instead of watering.

Foliar application: Fertilizers are applied to the foliage of the crops in the form of spray. This is followed where small quantities are required for normal growth of the crop. One percent urea solution spray is used in rice prior to flowering. Pulse crops are sprayed with two percent diammonium phosphate.

Deficiency of micronutrients is alleviated by foliar application. For example foliar spray of zinc sulphate at 0.5 percent, ferrous sulphate at 1.0 percent, boron at 0.2 percent.

Advantages

Foliar application improves fertilizer use efficiency since there is no transformation of nutrients. There will be quick response of crops. Followed when soil moisture is insufficient, where leaching losses are great, to alleviate deficiency and hidden hunger. Combined application of pesticides, fungicides and fertilizers reduce the cost of application.

Disadvantages

Large quantities cannot be applied and biurette content should be low in urea fertilizer.

Direct application to the soil: Anhydrous ammonia subjected to a pressure of 200 lbs/sq inch converts into liquid ammonia, which is then injected into the moist soil by using devices called injectors. Urine, sewage and cattle shed washings may be directly let into the field upon dilution with water.

Plant injections: Slanting holes are made into the tree trunks and micronutrient solutions are injected into the holes *e.g.*, zinc and iron fertilizers.

Application through irrigation water: Water soluble fertilizers are applied either through surface, subsurface or overhead irrigation. Gober gas slurry, liquid fertilizers, water soluble fertilizers dissolved in water and applied through irrigation. Application of fertilizers along with irrigation water is called fertigation.

Integrated nutrient management (INM)- Meaning and importance

Integrated nutrient management refers to supplementary as well as complementary use of organic manures, crop residues, green manures, biofertilizers and rural industrial wastes along with chemical fertilizers for meeting the nutrient demand of agricultural production systems.

Why integrated nutrient management?

- Single source of nutrients cannot meet the entire nutrient demand of the crop
- Continuous use of fertilizers alone results in decline in soil organic matter, imbalance in nutrient supply, decline in soil physical chemical and biological properties *viz*. soil structure, soil pH and soil microbial population.
- Integration of nutrient sources improves the nutrient use efficiency and sustains the soil productivity.
- Integrated nutrient management economize fertilizers without affecting the productivity.

Nutrient use efficiency: is defined as the unit quantity of grain (Kg grain) per unit (Kg) of applied nutrient. The efficiency of nutrient use varies with the nutrient. The efficiency of fertilizer nitrogen ranges from 40-50%, phosphorus from 15–20% while potassium 80%.

$$\text{Nitrogen use efficiency} = \frac{\text{Unit quantity grain/economic produce}}{\text{Unit quantity of N applied}}$$

Large quantity of applied nutrients are prone to various types of losses *viz.* flood water, leaching, denitrification, volatilization, immobilization, fixation or luxury consumption by plants.

The nutrient use efficiency can be increased by efficient fertilizer management practices to minimize the loss of nutrients and adoption of crop management practices (selection of variety and agronomic practices) to increase the crop yield.

Questions

I. Answer the following

1. What is optimum fertilizer dose?
2. What is the rationale of fertilizer selection to crops?
3. Why nitrogenous fertilizers are applied in split doses?
4. What is the meaning of basal dose and top dressing?
5. Why the entire dose of phosphorus and potassium fertilizers is applied as basal dose?
6. What are the disadvantages of broadcasting solid form of fertilizer in field crops?
7. What are the different methods of placement of fertilizers?
8. What are the advantages of integrated nutrient management?
9. What are the fertilizer management practices to increase nutrient use efficiency?
10. What are the crop management practices to increase nutrient use efficiency?

II. Define

1. Starter solution
2. Foliar fertilization
3. Fertigation
4. Integrated nutrient management
5. Nutrient use efficiency

III. Underline the correct answer

1. At the optimum fertilizer dose the crop produces
   (a) maximum economic yield
   (b) maximum yield
   (c) yield equivalent to state average
   (d) yield equivalent to the countries average
2. Phosphatic fertilizers suitable for plantation crops
   (a) citrate soluble and citrate insoluble phosphatic fertilizer
   (b) water soluble phosphatic fertilizer
   (c) citrate soluble phosphatic fertilizer
   (d) water soluble and citrate soluble phosphatic fertilizer
3. Sulphate of potash is appropriate for
   (a) rice (b) potato
   (c) tomato (d) sugarcane
4. Split application is followed for
   (a) phosphatic fertilizers (b) potassic fertilizers
   (c) nitrogenous fertilizers (d) micronutrient fertilizers
5. Elemental sulphur source should be incorporated into the soil at
   (a) three weeks before sowing
   (b) sowing
   (c) two weeks after sowing
   (d) three weeks after sowing
6. The number of top dressings recommended for sugarcane
   (a) 2 (b) 3 (c) 1 (d) 5
7. Broad casting of nitrogenous fertilizers is followed in
   (a) perennial forage crops (b) cotton
   (c) maize (d) sugarcane
8. Placement of NPK fertilizers reduces
   (a) fixation losses of phosphatic fertilizers
   (b) volatilization losses of nitrogen
   (c) run off losses of potassium
   (d) leaching losses of NPK
9. Starter solutions are used to
   (a) to provide the entire nutrient requirement of crop
   (b) nutrient required by the crop for initial growth
   (c) to minimize nutrient loss
   (d) to economize the cost of fertilizer

10. Nutrient deficiency or hidden hunger is alleviated by
    (a) foliar application of fertilizers
    (b) soil application of fertilizers
    (c) fertigation
    (d) use of starter solutions
11. Micronutrient deficiency is alleviated by
    (a) foliar application of micronutrient fertilizers
    (b) soil application of micronutrient fertilizers
    (c) fertigation
    (d) band placement of micronutrient fertilizers
12. The efficiency of nitrogenous fertilzers is
    (a) 40% (b) 20% (c) 80% (d) 60%
13. Water soluble phosphatic fertilzers are
    (a) band placed adjacent or below the seed row
    (b) broad casted on the soil surface
    (c) broadcasted and thoroughly incorporated
    (d) pocket placement

Chapter 11

Growth and Development of Crops

Plant growth is defined as the progressive development which is irreversible. It is expressed in terms of dry weight plants, height of plants, girth or diameter of the stem or tree trunks, spread of the tree canopy, number of branches or tillers. Plant growth is a function or various factors which may be expressed as $G = f(x, x_2, x_3 \ldots\ldots x_n)$.Some factors may be controlled while some cannot be controlled. The following are the factors which influence the plant growth.

Climate	Soil	Crop factors
Precipitation	Soil organic matter	Crop species
Air temperature	Texture	Variety
Relative humidity	Structure	Seed rate
Light	Cation exchange capacity of soil	Planting geometry
Altitude	Base saturation	Seed quality
Wind velocity	Slope and topography	Evapotranspiration
Carbon dioxide	Soil temperature	Water availability
Oxygen	Soil Management	Nutrition
Soil temperature	Tillage	Pests- insect pests, nematodes, snails
Soil aeration	Drainage	Diseases
Soil moisture	Depth of rooting zone	Weeds

Plant growth plotted against time assumes sigmoid or 'S' shaped curve. It is typical of growth pattern of individual organs of a whole plant or population of plants. It comprises different phases of growth which are illustrated in the figure given below.

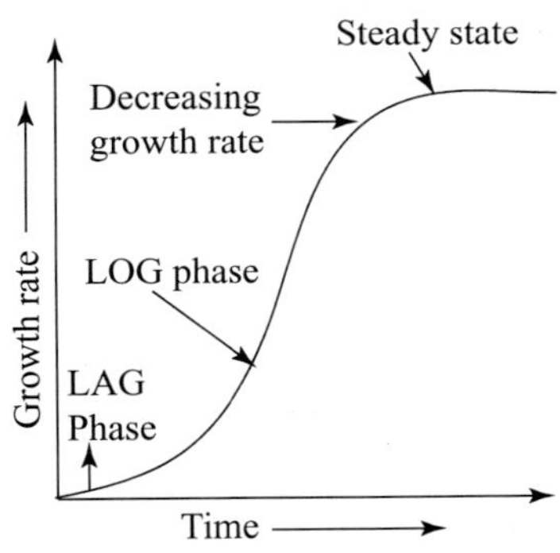

Fig.11.1 Sigmoid curve

Growth phases

1. Lag period of growth - during which internal changes occur that are preparatory to growth. The increase in size or weight is very slow or negligible during this period.
2. Log phase or log period of growth or the grand period of growth - during which growth is very fast. The logarithm of growth when plotted against time is a straight line.
3. Subsequently a phase in which growth rate gradually diminishes which is the decreasing growth period.
4. A point at which the plant reaches maturity and growth ceases which is the steady state.
5. Later, senescence and death of organism sets in, giving rise to another component of the growth curve- decreasing growth phase or senescence.

Developmental stages/Growth stages

Age of the plant is expressed as development stages. These developmental stages indicate age or progress towards maturity. The life cycle of the plant is divided into stages applicable for various crops are

(a) Germination
(b) Emergence
(c) Seedling
(d) Maximum vegetative growth stage
(e) Primordial differentiation
(f) Flowering
(g) Fruit maturity which comprise of Physiological maturity and Harvest maturity.

Germination and emergence

Seeds absorb moisture and puts forth radicle and plumule is known as germination. On subsequent growth the plumule is pushed through the soil and comes out of soil and is known as emergence. There must be optimum soil moisture, temperature and aeration and pH for germination of seeds. Poor germination and emergence of seedlings may be due to:

(a) Inadequate or excess moisture
(b) Cloddy soil surface.
(c) Hard crust on the soil surface
(d) Deep placement of seeds- if seeds are placed deep coleoptile may not be able to push the growing part out of the soil.
(e) Shallow placement or surface placement of seeds results in loss of seeds by bird picking.

Types of germination

1. Epigeal germination- Cotyledons emerge out of the soil by a curved elongation called hypocotyl. The cotyledons act as first leaves which provide food during the germination process. Later true leaves emerges which are protected by cotyledons. Epigeal germination is common in dicots.
2. Hypogeal germination - The cotyledons remain in the soil. The plumule grows which is pushed out by the elongation of epicotyl.

   Root development

   Radicle ⟶ Roots ⟶ Primary root
Secondary root
Seminal roots

At the end of seedling stage the lower nodes produce roots. These are nodal roots or adventitious roots. When seedling is transplanted they establish by producing adventitious roots. In dicots tap root system grows deeper into the soil with secondary or lateral roots. In monocots fibrous root system is seen.

Seedling stage

Subsequent to emergence, the plants putforth leaves and they no longer depend on seed reserves. During this stage, the plants establish by developing sufficient number of seminal roots and leaves. This stage ends when the plants putforth tillers and branches.

Maximum vegetative growth stage

During this stage, crops grow at a faster rate. Plants start producing branches or tillers and the crop covers the ground as much as possible to intercept more solar radiation. This stage is called as tillering stage in cereals and branching stage in non-cereals. During this stage there will be large accumulation of dry matter.

Primordial differentiation

This stage is called as panicle initiation in cereals and millets, squaring in cotton, flower bud initiation in sunflower. Plants enter into reproductive stage. The number of fruits or grains is decided during this stage. This stage is more sensitive to moisture and solar radiation in cereals.

Flowering stage

During this stage opening of flowers and anthesis takes place. It is also called blooming. This stage is very sensitive to moisture stress. Fruit set is decided based on the number of flowers that develop into fruits. Continuous rains cloudy weather, excess dew, desiccating winds affect pollination and seed set.

Fruit growth (Seed filling, Grain formation, Pod development)

During this stage photosynthates are translocated into the fruits. Fruit growth is also called as grain growth in cereals, pod growth in legumes, boll growth in cotton.

Fruit maturity

During maturity stage also there is translocation of photosynthates from leaves (source) to the storage organs (sinks). Biochemical reaction takes place in the fruits for the synthesis of proteins, oils, starches *etc.* from simple carbohydrates. The fruit maturity may be physiological maturity and harvest maturity. When the fruit growth is complete and no longer photosynthates are translocated to the fruits, it is known as physiological maturity. At physiological maturity the moisture percentage of grains is about 20 percent. After physiological maturity the crop is allowed for 10 or more days to reduce moisture percent and to provide time for development of late formed grains which is known as harvest maturity. During maturity stage soil should have optimum moisture for easy harvest of crops *viz.* groundnut, potato, sweet potato, radish, carrot, beetroot, onion and garlic whose economic product is underground. There is also need for cool and dry weather with bright sunshine during the maturity of crop.

Growth stages of crops

Cereals	Legumes	Oil seed crops	Fibre crops
Rice	Redgram	Sunflower	Cotton
Germination	Germination	Germination	Germination
Emergence	Emergence	Emergence	Seedling
Seedling	Seedling	Flower bud stage	Branching
Tillering	Flowering	Flowering stage	Boll formation
Lag vegetative growth	Pod formation	Grain formation	Square formation or squaring
Panicle initiation	Physiological maturity	Physiological maturity	Flowering
Panicle emergence	Harvest maturity	Harvest maturity	Boll formation
Flowering			Maturity
Grain formation			Harvest

Crop yield

1. Biological yield -The biomass yield excluding roots
2. Economic yield -That portion of the biological yield which has economic value

$$\text{Harvest index} = \frac{\text{Economic yield}}{\text{Biological yield}}$$

Plant growth substances: are biochemicals produced in plants (endogenous) or synthetic substances applied to plants (exogenous) which cause modification in plant growth and development.

Growth promoters: which enhance growth *e.g.*, Auxins, Cytokinins and Gibberellins.

Auxins: Stimulate elongation of cells of stems and coleoptiles. The endogenous auxins are indole acetic acid (IAA), phenol acetic acid. Some of the synthetic auxins are napthalene acetic acid (NAA), indole-3 - acetic acid (IAA), indole -3-propionic acid (IPA) and 2, 4-D.

Cytokinins: Stimulate cell division *e.g.*, Kinetin, Zeatin. Cytokinins promote orderly development of embryos of seed, break dormancy of seeds and buds. They delay senescence.

Gibberrellins: Promote cell elongation and increase in size of leaf, flower and fruit. Dormancy is broken and flowering is induced by gibberellins. *e.g.*, GA_3 (Gibberellic acid).

Growth retardants: which reduce growth *e.g.*, Abscisic acid and Ethylene
Abscisic acid (ABA): induces leaf and fruit abscission. Dormancy is induced due to accumulation of ABA. Under moisture stress ABA is produced in plants which facilitates closer of stomata and helps in maintaining turgidity of cells.

Ethylene: stimulates the swelling and isodiametric growth of stem and roots. Regulation of ethylene can trigger ripening of fruits or delay in ripening process.

Questions

I. Answer the following

1. What is plant growth? How is it expressed?
2. What are controllable and uncontrollable factors of plant growth?
3. Illustrate plant growth curve.
4. What are the phases of plant growth?
5. What are the developmental stages of crop plants?
6. What are the causes of poor germination and emergence of seedlings?
7. What are the types of germination?
8. What are the different types of roots in crop plants?
9. What is the significance of primordial differentiation stage?
10. What are growth promoters and growth retardants?
11. What is the role of
    (a) auxins (b) cytokinins (c) gibberellins
12. State the role of abscisic acid and ethylene.

13. List the growth stages of
    (a) rice (b) redgram
    (c) sunflower (d) cotton

II. Define

(a) Physiological maturity

(b) Harvest maturity

(c) Harvest index

(d) Biological yield

(e) Economic yield

III. Underline the correct answer

1. Growth plotted against time is a straight line during
   (a) lag phase of growth (b) log phase of growth
   (c) decreasing phase of growth (d) steady phase of growth
2. Epigeal germination is found in
   (a) rice (b) maize
   c) soyabean (d) wheat
3. Hypogeal germination is found in
   (a) redgram (b) cowpea
   (c) bengalgram (d) wheat
4. Squaring is the beginning of the reproductive stage in
   (a) chillies (b) cotton
   (c) rice (d) maize
5. The yield is decided during
   (a) primordial differentiation stage of the crop
   (b) flowering stage of the crop
   (c) fruit growth stage of the crop
   (d) maximum vegetative growth stage of the crop
6. Panicle initiation is one of the growth stage of
   (a) paddy (b) cowpea
   (c) sunflower (d) cotton
7. The ratio of economic yield to biological yield of the crop is
   (a) harvest index (b) net crop yield
   (c) gross crop yield (d) absolute yield
8. An example of growth retardant
   (a) naphthalene acetic acid (b) 2,4-D
   (c) abscisic acid (d) gibberlic acid
9. Harvest index of sugarcane is
   (a) 80 percent (b) 40–50 percent
   (c) 20–25 percent (d) 100 percent

Chapter 12

Ideotypes of Crop Plants

An ideotype is a crop plant having desirable morphological and physiological characteristics or both for cultivation in a specific agroclimate situation. An ideotype is constructed by the breeder on the basis of indepth study of morphological and physiological characters including their interrelationships, and intuition of the breeder based on his experience.

Ideotypes of cereals and pulses

Paddy

- Dwarf plants
- Erect flag leaf
- Acute leaf angle
- Leaves should remain green until maturity
- More number of tillers per hill
- Deep root system
- Resistance to drought in aerobic rice
- More number of grains per ear and test weight
- Uniform maturity of grains in ear
- Less number of chaffy grains
- Grains should be firmly attached to the rachis,
- High harvest index
- Tolerance/ resistance to pests and diseases
- Fast growing and covering the ground surface quickly
- Grain quality required for the purpose
- Fertilizer responsive- high N- use efficiency
- High water use efficiency

Wheat

- Dwarf plants
- More tillering and more number of effective tillers
- Upright or oblique leaves
- Leaves should remain green till maturity of the crop

- Husk should be easily detachable from the grain
  - Resistance to lodging
  - More number of grains per ear and test weight
  - Multiple resistance to diseases and pests
  - Harvest index should be more

Sorghum

- Dwarf, strong stem/thick stem
- Upright or oblique leaves, leaf area duration should be more
- Length of the ear head should be more
- More number of grains per ear and test weight
- Resistance or tolerance to pest and diseases
- Accumulate more photosynthates in the stem before the grain formation
- Deeper root system and less number of stomata
- Waxy coating
- Low hydrocyanic acid(HCN) content in fodder sorghum

Maize

- Dwarf
- Thick and strong stem
- Girth and length of the cob should be more
- More number of grains per cob
- Complete grain filling in the cob
- Larger leaf area and duration

Fodder Sorghum and Maize

- Tall and thin stem
- Fast growing
- Drought tolerant
- Sustain repeated cutting/ratooning

Ragi

- Strong stem
- Upright or oblique leaves
- More tillering
- Less stomata
- Deep root system
- Drought resistant

- High water use efficiency
- Open type earheads with long fingers
- Uniform maturity

Redgram

- Initial slow growth and long duration suitable for intercropping
- Fast growing and short duration suitable for late sowing and double cropping
- Erect acute angled branches
- More number of leaves and leaf area duration
- Synchronous flowering and maturity
- Pods nonshattering
- More number of pods and seeds per pod
- Drought resistance
- Pest and disease resistance
- Large seeded
- Nodes should be more

Chickpea

- More number of pods per plant
- More number of seeds per pod
- Seed size should be in accordance with the consumer preference
- Plant are designed to intercept more sunlight-erect plant types, leaf size and orientation to facilitate more light interception
- Leaf area duration should be more
- More number of nodes in the branches
- Profuse and deeper root system
- High harvest index
- A programmed phase of vegetative and reproductive growth to provide optimum balance of time and environmental resources
- Medium statured plants for intercropping and tall plants for mixed harvesting
- Short duration for shorter growing season and long duration types for longer growing season
- Multiple resistance to pests and diseases

Questions

I. Define

1. Define Ideotype
2. State the ideotype characters of

   (a) paddy (b) wheat
   (c) maize (d) ragi
   (e) redgram (f) chickpea

II. Underline the correct answer

1. Tall and thin stem is the ideotype character of

   (a) grain sorghum (b) fodder sorghum
   (c) sugarcane (d) bajra

2. Tiller number should be more in the ideotype of

   (a) maize (b) redgram
   (c) rice (d) sorghum

3. Accumulation of more photosynthates in the grain before grain formation is the ideal character of

   (a) sorghum (b) paddy
   (c) wheat (d) redgram

4. Desirable character of redgram cultivars suitable for intercropping

   (a) initial slow growth and long duration
   (b) initial fast growth
   (c) short duration
   (d) spreading branching types

5. Waxy coating is a desirable character of

   (a) sorghum (b) wheat
   (c) rice (d) maize

Chapter 13

Crop Adaptation and Distribution

Adaptation is any feature of an organism which has survival value under the existing condition of its habitat. The basic of successful crop production is the selection of an adapted plant species and varieties. Plants may have morphological adaptation or physiological adaptation or both for the existing environment.

Adaptation to soil moisture

Hydrophytes: More number of stomata on the upper surface of the leaf, aerenchyma tissue in the roots of rice plants, stem is hollow/straw, internodes are longer in floating and submerged rice,

Mesophytes: Stomata are normal. Present both on upper and lower leaf surface, medium root growth.

Xerophytes: Reduced leaf size, leaf number, number of stomata, deep root system, waxy coatings in sorghum, castor; leaves modified into spines *etc*. indeterminate spreading *e.g.*, groundnut, more tillering *e.g.*, bajra. Osmotic potential of the cell sap is high due to presence of hydrophilic colloids.

The adaptation of crop plants to climate, soil, and economic situation determine the desirability of growing any crop in a particular region.

Crop distribution

Principles of plant distribution: Environmental factors greatly influence the natural distribution of plants. Plant geographer Good (1931) formulated principles of plant distribution.

1. The plant distribution is primarily controlled by the distribution of climatic conditions.
2. The plant distribution is secondarily controlled by the distribution of edaphic factors.
3. The great movements of floras have taken place in the past and are still continuing (succession of floras as evidenced by the fossil records).
4. That the species movement (plant migration) is brought about by the transport of individual plants during their motile dispersal phases.

5. That there has been great variation and oscillation in climate, especially at higher latitudes, during the geological history of angiosperms.
6. That at least some, and probably considerable, variation has occurred in the relative distribution and outline of land and sea during the history of angiosperms.

The changes in the environment particularly climate resulted in plant movement and migration. This was explained by the Good's concept - specific tolerance according to which

1. Each and every plant species is able to exist and reproduce successfully only within a definite range of climatic and edaphic conditions. This range represents the tolerance of the species to external conditions.
2. The tolerance of a species is a specific character subject to the laws and process of organic evolution in the same ways as is its morphological characters, but the two are not necessarily linked.
3. Change in tolerance may or may not be accompanied by morphological change, and *vice versa.*
4. Morphologically similar species may show wide differences in tolerance and species with similar tolerance may show little morphological similarity. The relative distribution of species with similar tolerance is finally determined by the result of competition between them.
5. The range of tolerance of any larger taxonomic unit is the sum of (or total range or extent of) the ranges of tolerance of its constituent species.

Shelford (1913) proposed a general law of tolerance

1. Organisms with wide ranges of tolerance for all factors of the environment are likely to be widely distributed.
2. Organisms may have a wide range of tolerance for one factor and narrow range for another. When the conditions are not optimum for one factor the limit of tolerance may be reduced with respect to another factor.
3. The period of reproduction usually is critical when environmental factors are likely to be limiting.

Mason (1936) added that
The extremes of climatic conditions are more significant than the means, which emphasize the periphery, are limits of range where extremes of climatic factors are most likely to be limiting.

1. Both dispersal and establishment are essential for plant migration.
2. The tolerance theory should emphasize the factor function relationship.
3. During the life cycle of the plants certain critical phases have narrow tolerance range.

Cain (1944) further strengthened the tolerance theory by adding

1. That the biotic factors may be important,
2. That the environment is holocoenotic and
3. That the tolerance has a genetic basis

Biotic factors exert direct influences on plant distribution *e.g.*, Obligate insect pollination, seed dissemination, grazing by livestock.

Billings (1952) stated holocoenotic concept for plant distribution. He emphasized the importance of tolerance of plants to components of climatic, edaphic, and biotic factors for plant distribution.

1. Climatic- fire, water, wind pressure, atmospheric composition, cosmic radiation, solar radiation, terrestrial radiation and temperature.
2. Edaphic- soil, parent material, gravity, rotational forces, topography and geographic position *etc*.
3. Biotic–man, animals, other plants,

Limiting factors play a significant role in the distribution of plants.

- Law of Minimum (Liebig, 1840)–the growth of the plants is dependent on the amount of food stuff (or element) presented to it in minimal quantity.
- Taylor (1934) included environmental factors in addition to the nutrients. The most critical season of the year, the most critical year of the climatic cycle and the critical stages of development (germination, anthesis) .
- Blackman (1905) developed the "Theory of optima and limiting factors." -According to which "when a process was conditioned to its rapidity by a number of factors, the rate of the process was limited by the pace of the slowest factors".
- Lundegardh (1931) emphasized as a factor increases in intensity its relative effect on the plant growth decreases. This principle is sometimes called "Law of relativity".
- Livingston and Shreve (1921) proposed the concept of "Physiological limits" according to which for every vital function there is maximum and a minimum zero point with respect to any conditioning factor, beyond which the function ceases. Further for every distinct climatic area there appears to be a corresponding type of vegetation and this principle is probably of primary importance in the study of plant distribution.
- Odum (1969) attempted to combine these three theories. Plants appear to be controlled by three forces:

1. Quantity and variability of materials for which there are minimal requirements.

2. Physical factors which are critical.
3. Limits of tolerance of the plants themselves to these and other factors of the environment.

- Cain (1944) who noted that "The capacity of the species to tolerate or respond to its environment is governed by the laws of evolution and genetics, and the range of tolerance is the direct result of the diversity of the species".

Crop distribution

Climate plays dominant role in determining the distribution of world crops.

Tropical crops

Rubber, cocoa, banana, coffee, sugarcane are examples of crops adapted to tropical climate.

Sugarcane requires frost free climate with plenty of warm to hot sunny weather where monthly mean temperature ranges from 21-26°C throughout the year. The noble cane originating in Java is known especially for their narrow tolerance or sensitivity to environment. They have relatively thick soft stalks and high in sugar content. In Hawaii root temperature of 21°C is optimum for growth of cane and nutrient absorption. Below 10°C no growth takes place. It needs abundant sunshine and minimum temperature to which the crop tolerates is 12°C. At low elevations in Hawaii crop matures in two years or less. At 3000ft maturity is delayed. A growth in stalk length of 12ft per year is expected varying from 0.53 ft in February (mean temp of 21°C) to 2.14 ft in July (mean temp. of 25°C). Topography should be level. Rainfall of 112 cm is required without irrigation. Crop needs high NPK, the stalk and cane leaves incorporated add to a great deal of soil organic matter. Cuba endowed with a nearly ideal climate and favorable soil has been the world's leading producers of cane sugar. Other important production centers are Brazil, India, Puerto Rico, Hawaii, Australia, Peru, Argentina and Java. There are five species of sugarcane

- *Saccharum officinarum*- called noble cane, soft juicy and sweet found in Java
- *S. sinense*- hardy, slender with low sucrose known as Chinese or Uba cane
- *S. barberi*- slender stalks of Indian origin
- *S. spontanium*- Africa, Asia, Taiwan, Japan, New Guinea and Oceania. These are widely diverse forms which develop from rhizomes.
- *S. robustum*- stalks are large and woody- wild types found in New Guinea.

Crops of subtropics

Rice and cotton are the examples. Rice is the major crop of subtropics largely grown in China, India, Japan, Pakistan and Indonesia.According to Vavilov rice originated in India and spread eastwards into china by 3000BC. It may be considered as facultative hydrophyte. Plants can transport oxygen from leaves to the roots and thus live normally in an aquatic environment. It needs warm temperature throughout growing period 21°C or above. Silt loam with impervious subsoil is considered good for rice. Upland rice is grown without irrigation in regions of high rainfall.

Cotton is adapted to humid tropics where the frost free season is at least 200–210 days. Summer isotherms about 25°C, minimum rainfall requirement is 500–620 mm, needs abundant sunshine and moderate amount of moisture. Heavy rains at planting and during early stages are undesirable. Cool nights and cool days result in vegetative branch production while cool days and hot nights result in mainly fruiting branches. Well drained, alluvial, flood plains are best soils. Heavy clay soils promote more vegetative growth.

Cotton producing countries in world are united states (40 percent) followed by USSR, India, China, Egypt, Brazil, Pakistan, African highlands, Mexico, Turkey, Argentina and Peru. American upland cotton *Gossipium hirsutum* in southern Mexico and Central America, sea island type *G. barbadense* in the Andean region of Peru, Ecuador and Columbia. American cotton are remarkably plastic. *G. herbaceum* is called *desi* cotton or Asiatic cotton, *G. arborium* is called Egyptian cotton.

Crops of intermediate climate

Wheat, apple and maize are crops of intermediate climate.Wheat is spread in United States, China, USSR, Canada, India, Turkey, Italy, Argentina, France, Australia and Pakistan. In wheat carbohydrate and proteins are well balanced. It is adapted to wide variety of climates throughout the world. It is originated in Caucasus in Turkey–Iraq area and also in Afghanistan-West-Central -Asiatic area.

Tetraploid (AABB) and Hexaploid (AABBDD) resulted from earlier diploid (AA) wheat through hybridization and chromosome doubling. Man domesticated and influenced their spread which was quite different in their characteristics with those of wild forms. It is widely adapted throughout the world from near the equator to 60 ° N latitude and 40 ° S latitude. It is extensively grown in Steppe climate in United States, Canada and USSR. Best climate for wheat is cool moist growing season followed by a dry warm season for ripening. It is a dominant crop in Mediterranean climate with irrigation. It is poorly adapted to climates that are continually moist and hot.

Maize is next to wheat and rice in distribution and production in the world. Origin of maize is North America. There are different types of maize, pod corn, flour corn, popcorn, flint corn, sweet corn.

Maize is tropical in origin and is basically a short day plant. It is grown at 58 °N to 40 °S. it is grown in tropical, subtropical and intermediate climates. Temperature of maize belt varies from 21–24°C and the mean night temperature of 14°C. Night temperature should not be high. Maize has high efficiency of transpiration, but requires large quantities of water to produce high dry matter yield. Flowering and fertilization are more critical for moisture stress. It requires well drained soils with high nutrient status.

Questions

I. Define

(a) Crop adaptation
(b) Law of minimum
(c) Theory of optima and limiting factors
(d) Law of relativity
(e) Physiological limits

II. Answer the following

1. What are the adaptation mechanisms of
   (a) hydrophytes (b) mesophytes and (c) xerophytes?
2. What are the Good's Principles of plant distribution?
3. State the contributions of (a) Shelford (1913), (b) Cain (1944) and (c) Mason (1936) which supported the Good's principle of plant distribution.
4. What are the forces which control the plant distribution according to Odum (1969) ?
5. What are the climatic requirements of tropical crops, subtropical crops and the crops of intermediate climate.

III. Underline the correct answer

1. More number of stomata on the upper surface of the leaves is an adaptation mechanism of
   (a) hydrophytes (b) mesophytes
   (c) xerophytes (d) megatherms
2. High osmotic potential of the cell sap is an adaptation mechanism of the crop plants of
   (a) xerophytes (b) hydrophytes
   (c) mesophytes (d) floating plants
3. Principles of plant distribution were formulated by
   (a) Goods (1931) (b) Shelford (1913)
   (c) Mason (1336) (d) Cain (1944)

4. Theory of optima and limiting factors was developed by
   (a) Liebig (b) Taylor (1934)
   (c) Blackman (1905) (d) Lundegardh (1931)
5. Law of relativity was stated by
   (a) Lundegardh (1931) (b) Livingston and Shreve (1921)
   (c) Odum (1969) (d) Cain (1944)
6. A tropical crop
   (a) sugarcane (b) wheat
   (c) potato (d) beetroot
7. Subtropical crop
   (a) banana (b) sugarcane
   (c) rice (d) apple
8. A crop intermediate climate
   (a) maize (b) banana
   (c) coconut (d) rubber

Chapter 14

Principles of Cropping Systems

Cropping system is the cropping patterns used on a farm and their interaction with farm resources, other farm enterprises, available technology and environment (Physical, biological and socioeconomic) which determine their makeup (Krantz *et al.*, 1975). Cropping system is subsystem of the farming system.

The cropping system aims at increasing the benefits derived by crop production from available physical resources (land, climate, water resources *etc.*) that are not readily changed through optimization of input levels for a given crop or cropping pattern. The objective of any cropping system is to increase the efficiency of utilization of resources. The efficiency in measured by the quantity of product obtained per unit resource in a unit time in cropping system.

$Y = f(M.E)$

Y = yield

M = Management (type) and arrangement of crops, in time and space, choice of variety, agro techniques followed, pest and disease management.

E = Environment (soil, topography, climate, labour, capital, technology available, market, government policies *etc.*).

Cropping pattern is the yearly sequence and spatial arrangement of crops or of crops and fallow on a given area.

Types of cropping systems

Monocropping: It is the practice of growing only one crop on a piece of land in a year *e.g.*,

Finger millet in Eastern dry zone of Karnataka

Paddy in uplands of hilly zone of Karnataka

Redgram in North- eastern dry zone

Multiple cropping

Multiple cropping is intensification of cropping in time and space dimensions or growing two or more crops on the same field in a year.

Sequential cropping growing two or more crops in sequence on the same field in a farming year. The succeeding crop is planted after the preceding crop has been harvested.

Double cropping: Two crops are grown in a year *e.g.*,

Rice -Wheat

Rice - Rice

Paddy - Groundnut

Triple cropping - Three crops are grown in a year *e.g.*,

Rice - Rice- Cowpea

Finger millet - Potato - French bean

Quadruple cropping: Four crops are grown in a year *e.g.*,

Finger millet - Potato - Onion - French bean

Monoculture refers to growing single crop season after season or year after year without diversification *e.g.*, season after season over the years *e.g.*, rice-rice in Cauvery command area in southern dry zone and year after year groundnut is grown in central dry zone of Karnataka. It results in.

- Decline in soil fertility.
- Buildup of pest and disease causing organisms in the field.
- Risk of failure of crop and loss of farm income if there are adverse biotic or abiotic factors.
- Unsustainability of yield and income.

By monoculture the farmer attains specialization, inputs, implement and skill required are limited. Ease for management.

Crop rotation is the repetitive cultivation of an ordered succession of crops (or crops and fallow) on the same land. One cycle may take one or more years to complete.

Principles of crop rotation

1. Non leguminous crops should follow leguminous crops and *vice versa*

   Green gram - Wheat

   Maize - Soybean
2. Restorative crops should follow exhaustive crops

   Sunflower - Cowpea

   Maize - Groundnut
3. Fodder crops should be followed by field or vegetable crops

   Fodder sorghum - Groundnut/cabbage

4. Fouling crops should be followed by cleaning crops
   Cotton - Groundnut
5. Cleaning crops should be followed by nursery crops
   Potato - Ragi nursery
6. Deep rooted crops should be followed by shallow rooted crops
   Cotton - Green gram
7. Heavy irrigated crops should be followed by light irrigated crops.
   Paddy - Groundnut
8. Crops susceptible to pests/diseases/parasitic weeds should be followed by tolerant/break/trap crops/non host plants *e.g.*, Tomato/brinjal/Potato/ Tobacco - Sorghum/finger millet
9. Pasture/silage/hay/cleaning crops should be followed by grain crops.

Advantages of crop rotation

- Higher yields are obtained without additional cost.
- Maintenance or improvement in soil fertility, reduced soil erosion, protecting the soil against soil acidity and alkalinity.
- Reduced incidence of pests, diseases weeds.
- Improvement in soil health and crop quality.

Intercropping: Growing of two or more crops simultaneously on the same piece of land.

Row intercropping: Two or more crops are simultaneously grown on the same piece of land in definite row proportion.

Groundnut+ Redgram (3:1 or 8:2)

Maize + Soyabean (1:1)

Mixed intercropping/(Mixed cropping)

In this system seeds of different crops are mixed together and sown with no distinct row arrangement. Sometimes seeds may be broadcasted/sprinkled in an established crop.

Mixed row cropping

The seeds may be mixed and sown in definite rows. The seeds of different crops may be mixed and sown in a base crop as in 'Akkadi' system.

Relay intercropping

Seeds of the subsequent crop is sown or planted before the harvest of the previous crop. Intercrop competition is absent *e.g.*, Onion - Cotton

There will be efficient use of fertilizer, water and energy.

Strip intercropping

Growing two or more crops simultaneously in strips wide enough to permit independent cultivation but narrow enough for the crops to interact agronomically.

Groundnut + Maize
0 0 0 0 x x x x
0 0 0 0 x x x x
0 0 0 0 x x x x
00 0 0 x x x x
0 0 00 x x x x
0- Groundnut x- Maize

Multistoreyed cropping

The practice of growing different crops of varying height, rooting pattern and duration is called multistorey cropping. The vertical space is effectively utilized.

The leaf canopies of intercrop components occupy different vertical layers. The tallest component have foliage tolerant of strong light and high evaporative demand and the shorter component (s) with foliage requiring shade and relatively high humidity *e.g.*, Coconut + Pepper + Cocoa + Pineapple.

Interactions in InterCropping

1. **Indifferent:** The two species grown in a mixture do not interfere with each other.
2. **Competition:** When the resource *viz.* space, nutrients, water solar radiation requirement of the two species grown together is exactly same at a given time there will be competition. It may be mutual inhibition, mutual cooperation and compensation.
3. **Hampering effect**: The toxic secretions produced by one of the associated species will effect the growth of the other species. The toxic secretions are referred as allelo chemicals and the phenomenon is called 'Allelopathy'
4. **Stimulatory effect (negative allelopathy):** The productivity of one species is increased by some action of another species.

Complementary effect: When two crops are grown together is mixture the output of one crop increases while the output of the other crop also increases. It is termed as 'Annidation'.

Interactions in sequential cropping : Interactions may be positive or negative.

- System yield may increase when there is buildup of soil fertility due to addition of residues, nitrogen fixation and residual effect of fertilizers added.

- Yield of the succeeding crop declined when the proceeding crop exhaust available plant nutrients in soil.
- Poor crop stand due to destruction of soil physical properties when the preceeding crop spoils the soil structure *e.g.*, submerged paddy.
- There may be allelopathy or stimulatory effect on the crops grown in sequence *e.g.*, allelopathic effect of sunflower and stimulatory effect of leguminous crop.
- Yield loss due to buildup of pests, disease causing organisms, parasitic weeds when crops belonging to the same family are grown in sequence.

Questions

I. Define

(a) Cropping system
(b) Cropping pattern
(c) Monocropping
(d) Monoculture
(e) Sequential cropping
(f) Crop rotation
(g) Multistory cropping

II. Answer the following

1. What are the principles of crop rotation?
2. What are the different types of intercropping?
3. What are the interactions in intercropping?
4. What are the interactions in sequential cropping?

III. Underline the correct answer

1. Yearly sequence and special arrangement of crops or crops and fallow in a given area refers to
   (a) cropping system (b) cropping pattern
   (c) sequential cropping (d) multiple cropping
2. Decline in soil fertility and buildup of weeds, pests and diseases are associated with
   (a) monoculture (b) monocropping
   (c) inter cropping (d) relay cropping
3. Orderly succession of crops or crops and fallow on the same land is called
   (a) sequential cropping (b) intercropping
   (c) crop rotation (d) relay cropping

4. Intercrop competition is absent in
   (a) relay cropping (b) mixed intercropping
   (c) row intercropping (d) Akkadi cropping
5. Below ground and above ground resources are efficiently unitlized by growing crops having different rooting depth and height in
   (a) multiple cropping (b) multistorey cropping
   (c) row intercropping (d) strip intercropping
6. The component crop compensates the loss in yield of the associated crop in intercropping and the interaction is known as
   (a) indifferent (b) hampering effect
   (c) stimulatory effect (d) compensation
7. Cropping system adopted to control soil erosion is
   (a) strip cropping (b) relay cropping
   (c) mixed intercropping (d) multistorey cropping

Chapter 15

Crop Management Technologies in Problematic Areas

The problematic areas for crop production may be soils affected with acidity, high concentration of salts, water logged soils or soils in steep sloppy areas. These soils have limitations or ill effects for crop growth and productivity.

Acid soils

Strongly acidic soils pH < 5.0

Moderately acidic soils pH 5 to 6.5

Causes for formation of acid soils

- Carbon dioxide released during the respiration of micro organisms and plant roots react with water to form carbonic acid. The carbonic acid reacts with insoluble primary minerals present in the soil as well as the clay-humic micelle forming soluble bicarbonates.

$$\underset{\text{insoluble}}{CaCO_3} + H_2CO_3 \longrightarrow \underset{\text{soluble bicarbonate}}{Ca(HCO_3)_2}$$

- Similar reaction takes with magnesium, sodium and potassium to form bicarbonates of respective elements. In high rainfall area of the humid region soluble bicarbonates are lost by leaching.
- Addition of acid forming fertilizers- *e.g.*, urea, ammonium sulphate, ammonium chloride.
- Acid rain: Sulphur dioxide released during of the burning of fossil fuel *viz.* wood, coal, petroleum products reacts with water and forms H_2SO_4.

  $SO_2 + H_2O \rightarrow H_2SO_4$

  $NO + H_2O \rightarrow HNO_3$

  The acids so formed reach earth during rain in the form of acid rain.
- Removal of bases *viz.* calcium. and magnesium by the crops.

Ill-effects of acidic soil on plant growth

- Low availability of bases *viz.* calcium and magnesium.
- Toxicity of iron, manganese and aluminum.
- Low availability of molybdenum and boron.

- Availability of Phosphorus is reduced due to conversion of soluble phosphorus into insoluble form (fixation).
- Activity of bacteria decreases.

Management of acidic soils

- Application of agricultural lime *viz.* calcium oxide, calcium hydroxide depending on the soil acidity.
- Green manuring–either *in situ* green manuring or green leaf manuring.
- Liberal use of organic manures.
- Growing crops tolerant soil acidity.
- Avoid using acid forming fertilizers.

Management of Phosphorus in acid soils

- Avoid using water soluble phosphatic fertilizers. Use citrate soluble or insoluble phosphatic fertilizers *viz.* Rock phosphate, bonemeal.
- Placement of phosphatic fertilizers in band or spot application closer to plant roots, root dipping of seedlings with phosphate slurry.
- Submergence–growing paddy.
- Liberal use of organic manures, use of phosphorus solubilises *e.g., Aspergillus awamorii, Bacillus megterium.*

Advantages of liming in acidic soils

- Liming in acidic soil not only reduces acidity but also serves as source of calcium to plants which are deficit in acidic soil.
- Improves the activity of bacteria, nodulation in leguminous crops.
- Improves soil structure.
- Minimizes the harmful effect of soluble aluminium, iron and manganese.

Salt Affected Soils–Saline, Sodic and Saline–Sodic Soils

Causes for soil salinity

Salts accumulate in some surface soils of arid and semiarid regions because of insufficient rainfall to flush them from the upper soil layers. The salts are primarily chlorides, sulphates, carbonates or bicarbonates of calcium, magnesium and sodium. The sources of these salts are weathering of rocks and minerals, rainfall, groundwater and irrigation rich in salts. The salts are brought to or near the surface by upward movement of water which then evaporates, leaving the salts behind.

Saline soils

Electrical conductivity(EC) of the saturated extract of the soil is > 4 dS/m (deci siemens per meter) and exchangeable sodium percentage is < 15 and the pH is <8.5. Because the salts are neutral, the chlorides and sulphates of the base forming cations dominate. Saline soils have been called white alkali soils because of surface incrustation.

Effect of soil salinity on plant growth

Saline soils contain high concentration of neutral soluble salts sufficient to interfere seriously with the growth of the plants. Because of high osmotic potential of soil water than the internal water in the root cells, plants struggle to absorb water.

Germination and seedling growth will be poor.

Availability of micronutrients *viz.* iron, manganese, copper and zinc will decrease.

Absorption of potassium will be hindered due to high concentration of divalent cations *viz.* calcium and magnesium.

Management

- Leaching of salts by impounding water and flooding.
- Adequate provision should be made for drainage.
- Addition of green manure.
- Growing salt tolerant crops.
- Foliar application of micronutrients *viz.* zinc, iron and manganese if found deficient.

Sodic (Alkali) soils

Sodic soils do not contain any great amount of soluble salts. The detrimental effect of these soils on plants is not only due to the toxicity of Na^+, HCO_3 and OH^- but also reduced infiltration and aeration. ESP > 15, SAR > 13, Neutral soluble salts are low. An increase in OH^- ion results in increase in pH > 8.5, often rising it to 10 or higher. EC < 4dS/m. Sodic soils are called black alkali. Humus disperses on the surface of sodic soils causing black colour.

Effect of sodicity on plant growth

- High osmotic potential of soil solution hinders water uptake by the plants.
- Deficiency of nitrogen; sodium induced potassium deficiency, deficiency of zinc, iron, manganese and molybdenum.
- Germination and growth of seedlings will be adversely affected.

Management of sodic soils

- Sodic soils are reclaimed by application of gypsum

  $2NaHCO_3 + CaSO_4 \rightarrow CaCO_3 + Na_2SO_4 + CO_2$

  $2Na_2CO_3 + CaSO_4 \rightarrow CaCO_3 + Na_2SO_4$

Na_2SO_4 is leachable. Upon gypsum application salts are leached by flooding water.

- Addition of elemental sulphur or pyrites or sulphuric acid will also convert exchangeable sodium to sodium sulphate.
- Growing alkali resistant crops *e.g.*, rye barley, cotton, sweet clover, alfalfa
- Application farmyard manure or compost or green manuring (Dhaincha can tolerate sodicity).
- Use of acid forming fertilizers.

Saline-sodic soils

Contain appreciable quantities of neutral soluble salts and enough sodium ions to seriously affect most plants.

ESP > 15, EC >4, dS/m, pH 8.5 or less because of the presence of neutral salts, SAR > 13.

Management of saline-sodic soils

Leaching of saline-alkali soils will markedly raise the pH, unless calcium and magnesium salt concentrations are high in the soil or in the irrigation water. Increase in pH occurs because of removal of neutral salts. The exchangeable sodium readily hydrolyses and there by increases OH^- concentration. If the salts are removed pH rises > 8.5 and mineral colloids become dispersed, resulting in a tight impervious soil structure, toxicity of sodium to plants is due to increase in concentration of sodium.

Management

- Application of gypsum or sulphuric acid treated water.
- Application of greenmanure, farmyard manure or compost.
- Growing salt resistant crops.
- Management of nitrogen, potassium, and micronutrients zinc, iron and manganese.

Water logged soils

Poor drainage, unscientific irrigation, poor soil structure, and high water table are the causes for waterlogging.

Waterlogging reduces the choice of crops.

Management

- Growing crops tolerant to waterlogging *e.g.*, paddy, paragrass.
- Providing drainage.
- Growing trees which have high transpiration rate- biodrainage.

Steep sloppy lands

Water runoff, shallow soil depth due to continuous soil erosion and low soil fertility are the problems in the steep sloppy lands.

- Laying bunds and terraces according to the extent of land slope to minimize soil erosion.
- Contour cultivation–contour ploughing, sowing across the slope.
- Strip cropping, adoption of agroforestry, establishing perennial plantations *e.g.*, coffee, tea, cashew *etc.*
- Establishing live bunds, planting perennial plants having fibrous and strong root system.
- Mulching and application of organic manures.
- Avoid excessive cultivation.

Questions

I. Answer the following

1. How are the acid soils formed?
2. What are the management practices for acid soils?
3. What are the ill effects of acid soils on plant growth?
4. What are the phosphorus management practices for crops in acid soils?
5. What are the advantages of liming?
6. What are the characteristics of saline, sodic and saline- sodic soils?
7. How the saline soils are formed?
8. How do the saline soils affect the plant growth?
9. What are the management practices to alleviate the salinity of soils?
10. What are the management practices for reclaiming the sodic soils?
11. What are the management practices for crops grown in saline- sodic soils?
12. What are the management practices for crops grown in steep slopy lands?

II. Underline the correct answer

1. Gypsum amendment is used for reclamation of
   (a) acidic soils (b) saline soils
   (c) sodic soils (d) acidic soils
2. Lime is the amendment used to reclaim the soils and also as a sources of calcium
   (a) acidic soils (b) saline soils
   (c) sodic soils (d) saline sodic soils

3. Available iron and manganese is more in
   (a) acidic soils (b) saline soils
   (c) sodic soils (d) saline sodic soils
4. Citrate soluble or citrate insoluble phosphatic fertilizers are recommended for
   (a) saline soils (b) acidic soils
   (c) sodic soils (d) saline sodic soils
5. Phosphorus availability is more in
   (a) acidic soils (b) sodic soils
   (c) saline soils (d) drought affected soils
6. Sodic soils are deficient in
   (a) zinc (b) calcium
   (c) magnesium (d) phosphorus
7. Management option for sodium induced potassium deficiency in sodic soils
   (a) application of potassium
   (b) application of crop residues
   (c) adoption of management practices to achieve optimum Na /K ratio
   (d) split application of potassium
8. The management practices for steep sloppy lands include
   (a) mulching (b) strip cropping
   (c) contour cultivation (d) all the above stated a, b and c

Chapter 16

Harvesting and Threshing of Crops

Removal of entire plants or the economic parts after maturity from the field is called harvesting. Portion that is left behind on the field is known as stubble. The economic product may be grain, seed, leaf, root or entire plant.

Time of Harvesting

If the crop is harvested early, the produce contains high moisture and more immature grains. The yields will be low due to unfilled grains. It is very difficult to store the produce as shriveled grains with high moisture are highly prone to infestation by pest. The quality of the grain as well as germination percentage is also reduced. Late harvest results in shattering of grains, germination even before harvesting in rainy season and breakage during processing. Hence harvesting during correct time is very important.

Plants may be harvested at physiological maturity or harvest maturity. Plants are said to be in physiological maturity when the translocation of the photosynthates ceases to the economic part.

External symptoms of the physiological maturity of some field crops

Crop	Symptoms
Wheat	Complete loss of green colour from the glumes. Center spike are used as indicator grain
Barley	Loss of green colour from the glumes or peduncle
Maize	Black layer in the placental region of corn kernels
Sorghum	Black layer in the placental region of grain
Pearl millet	Appearance of bleached peduncle in some varieties
Soybean	Loss of green colour from the leaves
Red gram	Green pods turning brown, at about 25 days after flowering

Harvest maturity generally occurs seven days after physiological maturity. The important processes during this period are loss of moisture from the plants. The general symptoms of harvest maturity are yellowing of leaves, drying of grains or pods.

Method of Harvesting

Harvesting is done either manually or by mechanical means. In manual harvesting, sickle is the most important tool. The sickle has to be sharp, curved and serrated for efficient harvesting. The other tool used for harvesting is knife, especially the plant with a woody stem.

When there is labour shortage, or in peak period of labour demand when planting and harvesting coincides or for quicker harvesting, mechanical methods are adopted. Combine harvester or simply combine are used for harvesting.

Harvesting of crops is done by several ways depending on the crop. Cutting of the crop with sickle leaving small stubble near the ground is the common method of harvesting. This method is followed in crops whose maturity is almost uniform and occurs at a time. In indeterminate plants, harvesting is done at intervals as the economic product comes to maturity at different periods. Pods or fruits are picked at periodical intervals. Harvesting green gram, black gram, cotton *etc.* is known as picking and is done at 15 days interval.

Problems in Harvesting

Problems occur especially when it coincides with heavy rain or cyclones. The crops may be submerged and the seeds may start germinating on the plant itself. These problems are common in monsoon season especially rice in south India and can be overcome by growing dormant varieties.

When the economic part is underground, harvesting is difficult if the soil dries due to lack of rain. Running blade harrows may help to some extent to loosen the soil. Groundnut is harvested by running heavy blade harrows drawn by two pairs of bullocks. Tractor drawn sweep cultivators are also used to harvest groundnut crop.

Questions

I. Define

1. Physiological maturity
2. Harvest maturity

II. Answer the following

1. What is harvesting?
2. What is the significance of timely harvesting?
3. What are the symptoms of physiological maturity of wheat, maize, soyabean and sorghum?
4. What are the problems of harvesting?

III. Underline the correct answer

1. The best time of harvesting crop is
   (a) physiological maturity

(b) harvest maturity
(c) seven days earlier to harvest maturity
(d) seven days earlier to physiological maturity

2. Indeterminate crops are harvested by
   (a) picking matured pods
   (b) cutting close to the ground
   (c) harvest when all the pods are matured
   (d) harvest matured and immature pods or fruits and allow for ripening
3. Harvesting is done by picking in
   (a) tobacco (b) groundnut
   (c) cotton (d) paddy
4. Separation of grain from the straw is called
   (a) winnowing (b) threshing
   (c) harvesting (d) cleaning
5. Separation of grain from the husk is called
   (a) winnowing (b) grading
   (c) threshing (d) cleaning

Chapter 17

Weeds, Weed Management Herbicides and Allelopathy

Weeds are the plants growing where and when they are not desired. They compete with crops for soil moisture, nutrients, solar radiation and space, and thus reduce crop yields. Weeds also act as alternate hosts that harbour insect pests and diseases and other microorganisms. Some weeds release poisonous substances into the soil which may be harmful to crop plants, human and livestock. Weeds reduce the quality of marketable produce and render harvesting difficult, leading to increased expenditure on labours, equipment and chemicals for their removal.

Classification of weeds

A. Based on life cycle

Annuals–Complete their life cycle in a year or in a single growing season.

Based on the season they may be summer annuals and winter annuals.

Summer annual *e.g., Echinochloa crusgalli*

Winter annual *e.g., Phalaris minor*

Ephemerals–Short lived weeds e. g.*Phylanthus niruri, Ageratum conizoides.*

Biennials: Live longer than one year but less than two years *e.g., Raphanus raphanistrum*

Perennials- live longer than 2 years *e.g., Cynodon dactylon, Cyperus rotundus*

B. Based on Morphology

Grasses–*e.g., Echinocloa colonum*

Sedges–Belongs to cyperaceae family *e.g.,* Cyperus rotundus

Broad leaved weeds–dicotyledonous weeds *e.g., Celosia argentia, Digera arvensis* L., *Leucas aspera*

C. Based on habitat

Terrestrial and aquatic weeds. Aquatic weeds have been further categorized as floating, emerged and submerged weeds.

Floating weeds–Water hyacinth

Emergent weeds–Found along the shores of ponds and canals, *e.g.*, Typha.

Submerged weeds–Grow completely under water *e.g.*, Hydrilla.

D. Cotyledon character

Monocots–*Echinochloa colona*

Dicots - *Crotalaria Verucosa*

E. Based on origin

(a) Indigenous weeds (Apophytes) *e.g., Abutilon indicum*

(b) Introduced or exotic weeds or Alien weeds: *e.g., parthenium hysterophorus*

(c) Anthrophytes–Introduced by man *e.g., Phalaris minor*

F. Based on nature of stem

Presence of bark tissues on the stem and branches

Woody–These are shrubs also called brush weeds. *e.g., Lantana camera*

Semi- woody weeds *e.g., Croton sparsiflorus*

Herbaceous weeds *e.g., Amaranthus viridis*

G. Other groups

Noxious weeds/objectionable weeds are very difficult to control

1. Purple nut sedge–*Cyperus rotundus*
2. Bermuda grass–*Cynodon dactylon* L.
3. Jungle rice- *Echinocloa crusgalli* L.
4. Johnsongrass -*Sorghum halepense* L.
5. Water hyacinth- *Eichornia crassipes* L.
6. Quack grass- *Imperata cylindrica* L.
7. Lantana -*Lantana camera* L.
8. Parthenium–*Parthenium hysteroporus* L.
9. Baranyard grass–*Echinocloa colanum* L.
10. Field bind weed - *Convolvulus arvensis* L.

H. Satellite weeds/ mimicry weeds -The weeds have morphological similarity with crop plants.

Echinochloa crusgalli (Wild rice) in paddy, *Phalaris minor*–in wheat, Loranthus in tea.

I. Parasitic weeds

Weeds parasitize on specific host plants and deprive them of water, nutrients and assimilates.

Root parasite *e.g.*, Orobanche, Striga

Shoot parasite *e.g.*, Cuscuta, Loranthus

Partial root parasite *e.g.*, Striga in sorghum

Total root parasite *e.g.*, *Orobanche* in tobacco

Harmful effects of weeds

(a) Yield loss–The crop loss due to uncontrolled weeds vary from 15–80%. In general in farmer's field, yield loss varies from 15-30%. In some crops it is more than 50%. The combined effect of competition and allelochemicals is termed as *interference*.

(b) Quality–Quality of the crop is spoiled by admixture of weed seeds with economic produce *e.g.*, wild oat with wheat, oat, barley. In tea plantation loranthus leaves mix with tea leaves.

Cotton quality spoiled by dry weed fragments or interference with agricultural operations *e.g.*, *Cynodon dactylon*.

(c) Harbours pest and disease causing organisms *e.g.*,

The gram caterpillar of redgram–The alternate host is *Amaranthus sp.*

Black rust of wheat–The alternate host is A*gropyron repens*

Stem borer of rice –The alternate host is *Echinochloa, panicum*

Downy nuidew of maize –The alternate host is S*acharum spontaneum*

(d) Health hazards

Hay fever and asthama–Pollens of *Ambrosia artimissifolia*

Mexican poppy (*Argemone mexicana*) seeds crushed with mustard seeds for oil extraction cause blindness and death.

Allergy caused by *Parhenium hysterophorus*.

Characteristics of weeds

- Most of the weeds especially annuals produce enormous quantity of seeds *e.g.*, wild oats (*Avena fatua*) produces 250 seeds per plant, wild amaranth (*Amaranthus viridis*) produces nearly 11 million seeds per plant. Perennial weeds (Average of 61weeds) produce 16, 500 seeds per plant.
- Weeds have the capacity to withstand adverse conditions in the field (Evasiveness) because they can modify seed production and growth according to the availability of moisture and temperature. They have short period of plant growth, generally grow at faster rate and produce seed earlier than most of the crops growing in association (Ephemerals).
- Most of the weed seeds have dormancy remain viable for longer period *e.g.*, seeds of field bindweed (*Convolvulus arvensis*) viable for 50 years.
- Weed seeds have tremendous capacity to disperse through wind, water and

animals. Weed seeds mimic with the crop seeds due to their size and get transported from one place to another along with them.

- Vegetative reproduction–weeds propagate vegetatively by rhizomes, stolons, bulbs *etc*. This is a mechanism for weeds to withstand adverse conditions and help in persistence.
- Root system–Monocotyledonous weeds depend on seminal roots during seedling stage. Once nodal roots are developed at subsequent stage they are difficult to remove by hand weeding or intercultivation. Dicotyledonous weeds have tap root system. Tap root grows deeper into the soil, secondary thickening occurs and food materials are stored in the roots. At this stage even if the shoot is cut off the weeds may survive.
- Type of photosynthesis- Most of the weeds have C_4 type of photosynthesis. They deplete carbondioxide in the crop microclimate and can survive under moisture stress.

Crop weed competition

When crop seeds are planted weed seeds also germinate along with them. During the initial stages, both crop plants and weeds grow without competing with each other for nutrients, moisture and light. As the plants grow bigger and bigger, they vie each other for the resources and the one which is stronger numerically, morphologically, physiologically and biochemically will benefit more at the cost of other. The period of crop growth during which there is maximum impact of weeds on crop growth is termed as critical period of weed competition. Weeds must be kept at low level during the critical period of weed competition to ensure efficient use of resources and to attain maximum potential yield of crop.

The crop weed competition depends on weed flora, time of weed emergence, relative density of crop plants and weeds, and the crop factors *viz*. type of crops, varieties, ability to develop more roots, fast canopy covering, plant height, tillering/branching, canopy architecture *viz*.spreading or erect type.

Management factors which influence crop weed competition are depth of sowing, time of sowing, plant density, planting geometry, cropping systems adopted, method of fertilizer application and irrigation.

Hence the critical period of weed competition depends on crops and situation, crop varieties, weeds and management factors.

The critical period of weed competition of crops

1.	Rice (low land)	35 days
2.	Rice (upland)	60 days
3.	Sorghum	30 days

4.	Finger millet	35 days
5.	Pearl millet	35 days
6.	Maize	30 days
7.	Cotton	35 days
8.	Sugarcane	90 days
9.	Groundnut	45 days
10.	Soyabean	45 days
11.	Onion	60 days
12.	Tomato	30 days

Concepts of weed management

1. Control: Maintain the weed infestation as low as possible in order to prevent yield reduction and avoid further increase in weed infestation.
2. Control at economic threshold level is the point at which economic damage to a crop is about equal to the cost of controlling the weeds.

Principles of weed management

Prevention: Involves procedures that inhibit or delay establishment in areas that are not already inhabited by them. These practices restrict their introduction, propagation and spread of weeds on a local or a regional level. These include cultural practices such as seed cleaning. use of weed free seed, manure, and machinery, controlling weeds on field bunds and irrigation channels, screening irrigation water, restricting movement of farm animals, remove or destroy weeds before they attain reproductive stage or setting seeds, quarantine regulations to prevent entry of weed seeds/propagules from one country to another.

Eradication: Is the total elimination of a weed species from a field, area or region.

Control: Control practices reduce or suppress weeds but do not necessarily result in the elimination of any particular weed species. There are four general methods of weed control- physical, cultural, biological and chemical.

Physical methods: Hand pulling or hand weeding, hoeing, tillage, mowing, burning, sickling, flooding, smothering *etc.*, Physical energy through implements, either manually or bullock drawn or power operated.

Cultural methods: Quality seed with good germination, choosing varieties of crops well adapted to season, evolving quick-growing and short duration varieties of crop plants with larger leaf area and profuse branching/tillering, proper seed rate, depth, time and method of sowing, use of most appropriate method of irrigation and fertilization in a given cropping system. Close-row crops, crops or varieties having early and faster growth during cropping season (cowpea) fodder crops in general grow faster and thick sowing results in smothering of weeds. Crop rotations, intercropping with fast growing and slow growing crops (Redgram + cowpea, sugarcane + sunnhemp/ french bean/ soybean) mulching with crop residues/plastics. Stale seed bed technique.

Chemical methods: Use of chemicals generally referred as herbicides for the control of weeds is called chemical weed control.

Advantages: Can be adopted quite in time and in situations where manual and mechanical methods of weeding are difficult. It is easier and less time consuming. Pre emergent herbicides control weeds before their emergence.

Disadvantages: Requires technical knowledge, herbicides may pollute the soil and water, adverse effect on nontarget plants and soil microorganisms.

Classification of herbicides

A. Based on time of application

1. Preplant–before planting *e.g.*, fluchloralin
2. Pre emergence–after planting but before emergence *e.g.*, atrazine, pendimethalin.
3. Post emergence–After emergence of both crop and weeds *e.g.*, 2,4-D, Glyphosate.

B. Selectivity

Selectivity of herbicides: The unique feature of some herbicides to kill weeds selectively without any significant adverse effect on crops is termed as selectivity.

Hence different chemicals are used in different crops. Selectivity depends on dose, due to differences in morphological, anatomical, physiological and biochemical characteristics. Soil and climatic conditions may also alter uptake and metabolism of the herbicide.

1. Selective herbicides *e.g.*, atrazine in maize, isoproturon in wheat .Selective herbicides remove certain weeds in certain crops.
2. Non selective herbicides–Kill all vegetation including crops. *e.g.*, Paraquat, Glyophosate.

C. Based on the mode of action:

(a) Contact herbicide: Kill only the plant or portion of the plant that actually comes in contact with the herbicide. The herbicide to be effective must cover the total foliage. Selectivity depends on the differential wetting, difference with the form of plants and also on the placement of the spray. Selective and nonselective herbicides. *e.g.*, paraquat.

(b) Translocated herbicides: These herbicides move within the plant from the place of application to the place where they show their ultimate effect. *e.g.*, 2,4-D, Glyphosate.

D. Based on residual activity herbicides are classified as

(a) Residual herbicide–A residual herbicide maintains its phytotoxic effect in soil for more than 3-4 weeks *e.g.*, 2,4-D

(b) Non residual herbicide–These are inactivated in the soil within 3-4 weeks *e.g.*, paraquot.

Herbicide resistance in weeds

If the same herbicide, herbicides from the same chemical group, or those with the same mechanism of action is used on the same land for several successive years, development of resistance is more likely. The weeds may develop partial resistance, cross resistance (a weed biotype resistant to two or more herbicides due to single resistance mechanism) and multiple resistance (resistant plant possess two or more distinct resistance mechanisms) to applied herbicides.

Management of herbicide resistance in weeds

1. Crop rotation
2. Herbicide rotation
3. Herbicide mixtures–Combination of two or more herbicides
4. Integrated weed management.

Adjuvant: most of the herbicide molecules in their natural state are difficult to apply. They may be soluble powders, weltable powders or in emulsifiable form. To make them biologically effective adjuvants are added which increase their surface retention properties. Includes such materials as wetting agents, spreaders, adhesives, emulsifiers, dispersing agents and correctives

Biological weed control refers to use of living organisms *viz.* insects, disease causing organisms, herbivorous fish, snails or even competitive plants for the control of weeds.

Insects

Teleonemia scrupulosa (Lacewing bug) for control of *Lantana camera.*

Cactoblastis cactorum (*Dactylopius opuntiae)* for control of opuntia

Zygogramma biocolorata for control of *Parthenium hysterophorus.*

Disease causing plant pathogens: Divine–A liquid suspension of fungal spores of *Phytophthora palmivora* causes root rot, used to control Stranglevine (*Morrenia odorata*) in citrus.

Collego–Wettable powder contains fungal spores of *Colletotrichum gloeosporoides* causes leaf and stem blight, used to control joint vetch (*Aeschyomone virginica* L.) in rice and soybean.

Fish: Common/grasscarp(*Cyprimus carpio*) feed on grasses in aquatic ecosystem.

Plants: Cowpea in sorghum or redgram.

Advantages of biological weed control

- Least harm to environment
- No residual effect

Disadvantages of biological weed control

- Multiplication is costlier, slow control
- A few host specific bio agents are available.

Integrated weed management is the integration of different mentods *viz.* physical, cultural, chemical and biological means to keep weeds under check at an economic cost.

Advantage: efficient, cost of weed control is reduced, less dependence on herbicides and eco-friendly.

Allelopathy : It is the detrimental effect of chemicals or exudates produced by one plant species on the germination and growth or development of another plant species sharing the same habitat.

(a) Effect of weeds on crop plants: *e.g.*, tubers of *Cyperus esculentus* affect the drymatter production of maize. Seeds of wild oat affect the germination and early growth of wheat. Seeds of datura affect germination and growth of sunflower.

(b) Effect of crop plants on weeds: Root exudates of maize inhibit the growth of *Chenopodium album*.

(c) Effect of weeds on weeds: Leaf leachate of decaying leaves of polygonum contains floavonoides which are toxic to germination, root and hypocotyl growth of *Amaranthus spinosus.* Exudates of *Cassia sp.* affect the germination and growth of *Parthenium hysterophorus*.

Questions

I. Define

(a) Weeds
(b) Critical period of crop weed completion
(c) Herbicides
(d) Preplant herbicides
(e) Preemergence herbicides
(f) Post emergence herbicides
(g) Selective herbicides
(h) Contact herbicide
(i) Translocated herbicide
(j) Herbicide resistance in weeds
(k) Adjuvant
(l) Biological weed control
(m) Allelopathy
(n) Integrated weed management

II. Answer the following

1. Classify weeds based on
   (a) Life cycle (b) Morphology
   (c) Habitat (d) Origin
   (e) Nature of stem
2. What are the noxious weeds?
3. What are parasitic weeds?
4. What are harmful effects of weeds?
5. What are the characteristics of weeds?
6. What are the factors of crop weed competition?
7. State the critical period of weed competition of
   (a) Low land rice (b) Sorghum
   (c) Pearlmillet (d) Sugarcane
   (e) onion
8. State the concepts of weed management.
9. What are the principles of weed management?
10. What are the physical methods of weed control?
11. What are the cultural methods of weed control?
12. What are the advantages of chemical weed control?
13. What are the management options to break the acquisition of herbicide resistance in weeds?

III. Underline the correct answer

1. An example for winter annual
   (a) *Echinochloa crusgalli* (b) *Phalaris minor*
   (c) *Phylanthus niruri* (d) *Cynodon dactylon*
2. An example of a sedge
   (a) *Cynodon dactylon* (b) *Cyperus rotundus*
   (c) *Phalaris minor* (d) *Echinocloa crusgalli*
3. An example of ephemeral
   (a) *Avena fatua* (b) *Ehinochloa crusgalli*
   (c) *Phylanthus niruri* (d) *Argemone Mexicana*
4. Emergent weed
   (a) Typha (b) water hyacinth
   (c) Hydrilla (d) *Parthenium hysterophorus*
5. One of *the following is a Noxious weed*
   (a) *Echinochloa crusgalli* (b) Orabanche
   (c) Striga (d) Cuscuta
6. The crop having longer critical period of crop weed competition
   (a) rice (b) maize
   (c) finger millet (d) sugarcane
7. Preplant incorporation is followed for
   (a) fluchloralin (b) glyphosate
   (c) pendimethalin (d) 2,4-D
8. An example of selective herbicide
   (a) Atrazine (b) Glyphosate
   (c) Paraquat (d) Round up
9. A non selective herbicide
   (a) Glyphosate (b) Atrazine
   (c) pendimethalin (d) fluchloralin
10. One of the following is contact herbicide
    (a) 2,4-D (b) Paraquat
    (c) Glyphosate (d) Pendimethalin
11. *Lantana camera* is biologically controlled
    (a) *Cactoblasis cactorum* (b) *Zygogramma biocolarata*
    (c) Locusts (d) *Teleonemia scrupulosa*
12. Tubers of *Cyperusesculentus* affect the germination of
    (a) Maize (b) Wheat
    (c) Sunflower (d) finger millet
13. Exudates of cassia affect the germination of
    (a) *Parthenium hysterophorus* (b) *Chenopodium album*
    (c) *Avena fatua* (d) *Convolvulus arvensis*

References

Balasubramaniyan, P. and Sp. Palaniappan, 2001, Principles and Practices of Agronomy. Agrobios (India), Jodhpur 342002

Cheema, S.S., Dhaliwal, B.K. and Sahota, T.S. 1991 Agronomy–Theory and Digest.

Gopal Chandra De, 1995, Fundamentals of Agronomy, Oxford and IBH Publishing Co. Pvt. Ltd. New Delhi 110092

Gupta, O. P. 2010, Modern weed management, Agrobios, Jodhpur

ICAR, 2013, Hand book of Agriculture Indian Council of Agricultural Research, Krishi Anusandhan Bhavan I, Pusa NewDelhi–110012

ICAR, Agriculture in Ancient India

Reddy, S.R., 1999, Principles of Agronomy. Kalyani Publishers, New Delhi.

Sankaran, S. and Subbiahmuduliar, V. T. 1991, Principles of Agronomy, The Bangalore printing and publishing Co. Ltd. Mysore Road, Bengaluru 560018

Singh, S.S., Principles and Practices of Agronomy, 1998, Kalyani Publishers, 1/1 Rajinder Nagar Ludhiana -141008

Tisdale, S.L., Nelson, W.L., Beaton, J.D. and Havlin, J.L., 2002 Soil fertility and Fertilizers. Prentice Hall of India Pvt. Ltd.New Delhi 11001

Wilsie, C.P. 1974, Crop Adaptation and Distribution.

Yellamanda Reddy T. and Sankara Reddy, G.H., 2012, Principles of Agronomy, Kalyani Publishers, 1/1 Rajinder Nagar Ludhiana -141008

Annexure I

Common and Scientific Names of Field Crops

Cereals

Rice	:	*Oryza sativa* L.
Wheat	:	*Triticum* spp.
Maize	:	*Zea mays* Linn.
Sorghum	:	Sorghum spp.
Pear millet	:	*Pennisetum typhoides* Stapf and Hubb.
Finger millet	:	*Eleusine coracana* Gaertn.
Italian millet	:	*Setaria italica* Beauv.
Little millet	:	*Panicum miliare* Linn.
Common millet	:	*Panicum miliaceum* Linn.
Barnyard millet	:	*Echinochloa colona* var. Frumentacea Fisher
Kodo millet	:	*Paspalum scrobiculatum* Linn.
Oats	:	*Avena* spp.
Barley	:	*Hordeum vulgare.*
Rye	:	Secale cereal
Pseudo-cereals	:	Grain Amaranths–*Amaranthus* spp. Quinoa - *Chenopodium quiona* Buck wheat - *Fagopyrum esculentum* Moench.

Pulses

Redgram or Pigeon pea	:	*Cajanaus cajana* Mill sp.
Bengalgram or Chickpea	:	*Cicer arietinum* Linn.

The Phaseolus Group of Pulses

Greengram	:	*Phaseolus aureus* Roxb.
Blackgram	:	*Phaseolus mungo* Linn.
Pillipesara	:	*Phaseolus trilobus* Ait
Dew gram	:	*Phaseolus aconitifolius* Jacq.
Teegapesara	:	*Phaseolus sublobatus* Roxb.

Lima bean	:	*Phaseolus lunatus* Linn.
French bean	:	*Phaseolus vulgaris* Linn.
Scarlet Runner	:	*Phaseolus multiflorus* Wild.

The Dolichos Group of Pulses

Common bean	:	*Dolichos lablab* Linn.
Lablab bean	:	*Dolichos lablab* L. var. Typicus Prain.
Field bean	:	*Dolichos lablab* L. var. Lignosus Prain.
Horsegram	:	*Dolichos biflorus* Linn.
Cowpea	:	*Vigna sinensis* L. Savi.
Peas	:	*Pisum sativum* Linn.
Chickling vetch or Grasspea	:	*Lathyrus sativus* Linn.
Lentil	:	*Lens esculenta* Moench.
Soybean	:	*Glycine max* (L.) Merr.
Gaur	:	*Cyamopsis tetragonoloba* (L.) Taub.

Sugars and Starches

Sugarcane	:	*Saccharum* spp.
Sugarbeet	:	*Beta vulgaris* L.
Potato	:	*Solanum tuberosum* Linn.
Tapioca	:	*Manihot esculenta* Crantz.
Sweet Potato	:	*Impomoea batatus* Lam.
West Indian Arrow-root	:	*Maranta arundinacea* Linn.
East Indian Arrow-root	:	*Curcuma angustifolia* Roxb.
Queensland Arrow-root	:	*Canna edulis* Ker-Gawl.
Buckwheat	:	*Fagopyrum esculentum* Moench.
Sago Palm	:	*Mketroxylon sagu* Rottl.
Yams	:	*Dioscorea* spp.
Taro	:	*Colocasia esculenta* (L.) Schott.
Coco yams	:	*Xanthosoma*s Spp.

Oils and Fats

Groundnut or Peanut	:	*Arachis hypogaea* Linn.
Coconut	:	*Cocos nucifera* Linn.
Castor	:	*Ricinus communis* Linn.

Gingelly	:	*Sesamum indicum* Linn.
Rape and Mustard	:	*Brassica* spp.
Safflower	:	*Carthamus tinctorius* Linn.
Niger	:	*Guizotia abyssinica* Cass.
Sunflower	:	*Helianthus annuus* Linn.

Fibre Crops

Cotton	:	*Gossypium* sp.
Jute	:	*Corchorus* spp.
Kenaf Fibre	:	*Hibiscus cannabinus* Linn.
Roselle	:	*Hibiscus sabdariffa* Linn.
Flax	:	*Linum usitatissimum* Linn.

Fumitories and Masticatories

Tobacco	:	*Nicotiana* spp.
Betel Vine	:	*Piper betle* Linn.
Arecanut	:	*Areca catechu* Linn.

Beverages and Rubber

Tea	:	*Camellia sinensis* Kuntz.
Coffee	:	*Coffea* spp.
Cocoa	:	*Theobroma cocao* Linn.
Chicory	:	*Cichorium intybus* Linn.
Para Rubber	:	*Hevea brasiliensis* Muell. Arg.

Green Manures and Green Leaf Manures

Green Manure Crops	:	*Sesbania speciosa* Taub.
Dhaincha	:	*Sesbania aculeate* Poir.
Sunnhemp	:	*Crotalaria juncea* Linn.
Wild Indigo	:	*Tephrosia purpurea* Pers.
		Tephrosia noctiflora
Bengal Indigo	:	*Indigofera tinctoria* Linn.
Pillipesera	:	*Phaseolus trilobus* Ait.
Cluster beans	:	*Cyamopsis tetragonoloba* (L.) Taub.
Green Leaf Manure Plants	:	*Gliricidia sepium* Steud.
		Ipomoea carnea Jacq.
Tanners' Cassia	:	*Cassia auriculata* Linn.

Forage Crops

Guinea grass	:	*Panicum maximum* Jacq.
Napier Grass	:	*Pennisetum purpureum* Schum.
Napier Hybrids	:	
Para Grass	:	*Brachiaria mutica* Stapf.
Teosinte	:	*Euchlaena Mexicana* Shrad.
Buffel Grass	:	*Cenchrus* spp.
Marvel Grass	:	*Dicanthium annulatum* Stapf.
Maxican penicgrass	:	*Panicum antiodotale* Retz.
Spear Grass	:	*Heteropogon contortus* Beauv.
Musal grass	:	*Iseile malaxum* Hack. :
Lucerne (Alfalfa)	:	*Medicago sativa* Linn.
Sunnhemp	:	*Crotalaria juncea* Linn.
Ramie	:	*Boehmeria nivea* Gaud.
Hemp or Soft Hemp	:	*Cannabis sativa* Linn.
Agave fibres	:	*Agave sisalana; Agave cantala; Agave vera-cruz;* *Agave wightii: Agave furcroydes; Furcroea gigantea; Sansevieria spp.*
Other Fibres	:	*Luffa cylindrica; Coir; Borassus flabellifer: Musa textilis; Ceiba pentandra; Bombax malabaricum*

Spices and Condiments

Pepper	:	*Piper nigrum* Linn.
Chillies	:	*Capsicum annum* Linn.
Ginger	:	*Zingiber officinale* Roscoe
Turmeric	:	*Curcuma longa* Linn.
Onion	:	*Allium cepa* Linn.
Garlic	:	*Allium sativum* Linn.
Coriander	:	*Coriandrum sativum* L.
Cumin	:	*Cuminum cyminum* L.
Fennel	:	*Foeniculum vulgar* Mill.
Vanilla	:	*Vanilla planifolia* Andr.
Dill	:	*Peucedanum graveolens* Linn.

Fenugreek	:	*Trigonella foenum-graecum* Linn.
Caraway	:	*Carumcarvi* Linn.
Anise	:	*Pimpinella anisum* Linn.
Saffron	:	*Crocus sativus* Linn.
Asafoetida	:	*Ferula asafoetida* Linn.
Cardamom	:	*Elettaria cardamomum* Maton.
Nutmeg and Mace	:	*Myristica fragrans* Houtt.
Cloves	:	*Eugenia aromatica* (Linn.) Baill
Cinnamon	:	*Cinnamomum zeylanicum* Breyn.

Annexure II

International and National Agricultural Research Centres/Institutes

1. CGIAR	Consultative Group on International Agricultural Research. Supports International Research Institutes through funding.
2. CIP	Centre International de la papa–Lima, Peru, Potato
3. IBPGR	International Board for Plant Genetic Resources–Rome, Italy–On Plant Genetic Resources.
4. ICARDA	International Centre for Agriculture Research in Dry Areas, Aleppo-Syria Dry farming, Wheat, Barley, Broadbeans, Lentil, Chickpea, Forage Crops.
5. ICRISAT	International Crop Research Institute for Semi Arid Tropics, Hyderabad, India. Chickpea, Pigeonpea, Pearlmillet, Sorghum, Groundnut, Semi arid farming systems.
6. ILCA	International Livestock Centre for Africa–Addis Ababa, Ethiopia. Livestock production systems.
7. ILARD	International Laboratory for Research on Animal Diseases, Nairobi, Kenya.
8. IRRI	International Rice Research Institute, Los Banos, Manila, Philippine–Rice. Rice based cropping systems.
9. IITA	International Institute for Tropical Agriculture Ibadan, Nigeria, Limabean, Soybean, Maize, Rice, Sweet potato, Yams, Cassava, Cowpea.
10. IFPRI	International Food Policy Research Institute, Washington, USA–Food Policy.
11. ISNAR	International Service for National Agricultural Research–The Hague, Netherlands–On National Agricultural, Monnovia, Liberia, Rice in West Africa.
12. WARDA	West African Rice Development Association, Monnovia, Liberia, Rice in West Africa.
13. ICLARM	International Centre for Living Aquatic Resources Management, Manila, Philippines.
14. IIWM	International Institute for Water Management, Colombo, Srilanka.
15. CIAT	Centro International de Agricultural Tropical–Cali, Colubia on Cassava, Field Bean, Rice, Tropical.

16. CIMMYT Centro International de Mejormiento Maize Y trigo–Mexico city Mexico on Maize, Wheat, Barley and Triticale.

ICAR Institutes, National Bureaux and National Research Centres

Institutes

National Research Institutes

1. Indian Agricultural Research Institute, New Delhi 110012.
2. Indian Veterinary Research Institute, Izathnagar 243122 (Uttar Pradesh)
3. National Dairy Research Institute, Karnal 132001.
4. Central Institute of Fisheries Education, Bombay 400061.

Crop Sciences Institutes

5. Central Institute of Cotton Research Nagpur 440010 (Maharashtra)
6. Central Research Institute for Jute and Allied Fibres. Barrackpore 743101 (West Bengal)
7. Central Rice Research Institute, Cuttack 753006 (Orissa)
8. Central Tobacco Research Institute Rajahmundry 533101 (Andhra Pradesh)
9. Indian Grassland and Fodder Research Institute, Jhansi 284003 (Uttar Pradesh)
10. Indian Institute of Pulses Research, Kanpur 208024 (Uttar Pradesh)
11. Indian Institute of Sugarcane Research Lucknow 226002 (Uttar Pradesh)
12. Sugarcane Breeding Institute, Coimbatore 641007 (Tamil Nadu)
13. Vivekananada Parvatiya Krishi Anusandhan Shala, Almora 263601 (Uttar Pradesh)
14. Directorate of Rice Research, Hyderabad 500030 (Andhra Pradesh)
15. Project Directorate–Biological Control, Bangalore 560024 (Karnataka)
16. Project Directorate–Oilseeds, Hyderabad 500030 (Andhra Pradesh)
17. Project Directorate–Wheat, Karnal 132001 (Haryana)
18. Project Directorate–Maize, IARI, New Delhi
19. Project Directorate–Sorghum, Hyderabad 500030 (Andhra Pradesh)

Horticulture and Plantation Crops Institutes

20. Central Institute of Horticulture for Northern Plains, Lucknow 226006 (Uttar Pradesh)
21. Central Tuber Crops Research Institute, Trivandrum 695017 (Kerala)
22. Central Plantation Crops Research Institute, Kasargod 6701224 (Kerala)
23. Central Potato Research Institute, Shimla 171001 (Himachal Pradesh)

24. ICAR Research Complex for Goa, Goa Ela Old 403402
25. Indian Institute of Horticultural Research, Bangalore 560080 (Karnataka)
26. Central Institute for Temperate Horticulture, Srinagar 190001 (J & K)
27. Project Directorate–Vegetable Research, Varanasi 221001 (Uttar Pradesh)

Resources Management Institutes

28. Central Agricultural Research Institute for Andaman and Nicobar Group of Islands, Port Blair 744101 (Andaman)
29. Central Soil and Water Conservation Research and Training Institute, Jodhpur 432003 (Rajashthan)
30. Central Research Institute for Dryland Agriculture, Saidabad, Hyderabad 500659
31. Central Soil and Water Conservation Research Institute, Dehra Dun 248195 (Uttar Pradesh)
32. ICAR Research Complex for NEH Region, Barapani 793003 (Meghalaya)
33. Indian Institute of Soil Sciences, Bhopal–462011 (Madhya Pradesh)
34. Directorate of Water Managment, Rahuri 413722 (Maharashtra)
35. Project Directorate for Cropping Systems Research, Modipuram 250110 (Uttar Pradesh)

Agricultural Engineering Institutes

36. Central Institute of Agricultural Engineering, Bhopal–462003 (Madhya Pradesh)
37. Central Institute of Post-harvest Engineering and Technology, PAU Campus, Ludhiana 141004, (Punjab)
38. Central Institute for Research on Cotton Technology, Bombay 400019 (Maharashtra)
39. Indian Lac Research Institute, Namkum, Ranchi 834010 (Bihar)
40. Jute Technological Research Laboratory, Calcutta 700040 (West Bengal)

Animal Sciences Institutes

41. Central Avian Research Institute, Izatnagar 243122 (Uttar Pradesh)
42. Central Institute for Research on Buffalo, Hissar 125001 (Haryana)
43. Central Institute for Research on Goat Makhdoom 281122 (Uttar Pradesh)
44. Central Sheep and Wool Research Institute, Avikanagar 304501 (Rajasthan)
45. National Institute of Animal Genetics, NDRI Campus, Karnal 132001 (Haryana)
46. National Institute of Animal nutrition and Physiology.

47. High Security Animal Disease Laboratory, Bhopal.
48. Project Directorate on Cattle, Modipuram, Meerut 250110 (Uttar Pradesh)
49. Project Directorate on Poultry, Rajendranagar, Hyderabad 500030 (Andhra Pradesh)

Fisheries Institutes

50. Central Institute of Brackishwater Aquacluture, Madras 600008 (Tamil Nadu)
51. Central Inland Capture Fisheries Research Institute, Barrackpore 743101 (West Bengal)
52. Central Institute of Freshwater Aquaculture, Dhauli, Bhubaneshwar 751002 (Orissa)
53. Central Institute of Fisheries Technology, Cochin 682029 (Kerala)
54. Central Marine Fisheries Research Institute, Cochin 682029 (Kerala)

Economics and Statistical Institute

55. Indian Agricultural Statistics Research Institute, New Delhi 110012.

Research Management

56. National Academy of Agricultural Research Management, Rajendranagar, Hyderabad 500 030 (Andhra Pradesh)

National Bureaux

57. National Bureau of Plant Genetic Resources, Pusa, New Delhi 110012
58. National Bureau of Soil Survey and Land-use Planning, Nagpur 440006 (Maharashtra)
59. National Bureau of fish Genetic Resources, Allahabad 211002 (Uttar Pradesh)
60. National Bureau of Animal Genetic Resources, NDRI Campus, Karnal 132 -001 (Haryana)

NATIONAL RESEACH CENTRES

Crop Sciences

1. National Research Centre for Integrated Pest Management, Faridabad 121001 (Haryana)
2. National Research Centre for Groundnut, Junagadh 362002
3. National Research Centre for Sorghum, Hyderabad 500030 (Andhra Prdesh)
4. National Research Centre for Soybean, Indore 462001 (Madhya Pradesh)
5. National Research Centre for Biotechnology, New Delhi 110012.

6. National Research Centre for Rapeseed and Mustard, P.O. Box No. 41, Bharatpur (Rajasthan)

Horticulture

7. National Research Centre for Cashew, Puttur 574202 (Karnataka)
8. National Research Centre for Citrus, Nagpur 440006 (Maharashtra)
9. National Centre for Mushroom Research and Training, Solan 173213 (Himachal Pradesh)
10. National Research Centre for Spices, Calicut 673012 (Kerala)
11. National Research Centre for Arid Horticulture, C-35, Sadul Ganj, Bikaner 334001 (Rajasthan)
12. National Research Centre for Banana, Podavur, Tiruchirapalli, Tamil Nadu
13. National Research Centre for Onion and Garlic, Nasik (Maharashtra)
14. National Research Centre for Orchids
15. National Research Centre for Medicinal and Aromatic Plants, Boravi Seed Farm, Boravi 387310, Dist. Kheda, Gujarat
16. National Research Centre for Grapes, Pune.
17. National Research Centre for Oil Palm

Resource Management

18. National Research Centre for Agroforestry, IGFRI, Jhansi 284003 (Uttar Pradesh)
19. National Research Centre for Weed Science, Jabalpur 482004 (Madhya Pradesh)
20. Water Technology Centre for Eastern Region, Bhubaneswar 751016 (Orissa)

Animal Science

21. National Research Centre for Camel, Jorbeer, Bikaner 334001 (Rajasthan)
22. National Research Centre for Equines, Hisar 125001 (Haryana)
23. National Research Centre for Meat and Meat Products, Izatnagar 243122 (Uttar Pradesh)
24. National Research Centre for Mithun, Jhamapani 797106 (Nagaland)
25. National Research Centre for Yak Nikamadang 790101 (Arunachal Pradesh)
26. National Biological Centre on Animal Health, IVRI, Izatnagar 243122 (Uttar Pradesh)

27. National Biological Centre on Animal Production, NDRI Campus, Karnal 132001 (Haryana)

Fisheries

28. National Research Centre on Coldwater Fisheries, Haldwani 263139 (Uttar Pradesh) Economics and Statistics
29. National Research Centre of Agricultural Economics and Policy Research, IASRI Campus, New Delhi 110012.

Agri. Extension/ Agri. Engineering

30. National Research Centre for Women in Agriculture, Bhubaneswar 751001 (Orissa)

Annexure III

List of Agricultural Universities

Central Agricultural Universities

1. Central Agricultural University,P.O. Box 23, Imphal-795004, Manipur
2. Rani Laxmi Bai Central Agricultural University, Jhansi, Uttar Pradesh
3. Dr. Rajendra Prasad Central Agricultural University, Pusa (Samastipur)

State Agricultural Universities

Andhra Pradesh

1. Acharya NG Ranga Agricultural University, Guntur
2. Dr. YSRHU (APHU), Venkataramannagudem
3. Sri Venkateswara Veterinary University, Tirupati

Assam

4. Assam Agricultural University, Jorhat

Bihar

5. Bihar Agricultural University, Sabour, Bhagalpur
6. Bihar Animal Sciences University, Patna

Chhattisgarh

7. Indira Gandhi Krishi Viswa Vidhyalaya, Raipur
8. Chhattisgarh Kamdhenu Visva Vidyalaya, Durg

Gujarat

9. Sardar Krushinagar Dantiwada Agricultural University, Dantiwada
10. Anand Agricultural University, Anand
11. Navsari Agricultural University, Navsari
12. Junagarh Agricultural University, Junagarh
13. Kamdhenu University, Gandhinagar

Haryana

14. Chaudhary Charan Singh Haryana Agricultural University, Hisar
15. Lala Lajpat Rai University of Veterinary & Animal Sciences, Hisar
16. Haryana State University of Horticultural Sciences, Karnal

Himachal Pradesh

17. Ch. Sarwan Kumar Himachal Pradesh Krishi Viswa Vidyalaya, Palampur
18. Dr. Yaswant Singh Parmar University of Horticulture & Forestry, Solan

Jharkhand

19. Birsa Agricultural University, Ranchi

Jammu & Kashmir

20. Sher-e-Kashmir University of Agricultural Science & Technology, Srinagar
21. Sher-e-Kashmir University of Agricultural Science & Technology, Jammu

Karnataka

22. University of Agricultural Sciences, Bangalore
23. Karnataka Veterinary, Animal and Fisheries Sciences University, Bidar
24. University of Agricultural Sciences, Raichur
25. University of Agricultural Sciences, Dharwad
26. University of Horticulture Science, Bagalkot
27. University of Agriculture & Horticulture Sciences, Shivamogga

Kerala

28. Kerala Agricultural University, Thrissur
29. Kerala University of Fisheries and Ocean Studies, Panangad, Kochi
30. Kerala Veterinary and Animal Sciences University, Pookode, Wayanand, Kerala

Madhya Pradesh

31. Rajmata Vijayaraje Scindia Krishi Vishwa Vidyalaya, Gwalior
32. Nanaji Deshmukh Pashu Chikitsa Visva Vidyalaya, Jabalpur
33. Jawaharlal Nehru Krishi Viswa Vidyalaya, Jabalpur

Maharashtra

34. Dr. Balasaheb Sawant Kokan Krishi Vidyapeeth, Dapoli
35. Maharastra Animal & Fisheries. Sciences University, Nagpur
36. Vasantrao Naik Marathwada Krishi Vidyapeeth, Parbhani
37. Mahatama Phule Krishi Vidyapeeth, Rahuri
38. Dr. Punjabrao Deshmukh Krishi Viswa Vidyalaya, Akola

Orissa

39. Orissa University of Agriculture & Technology, Bhubaneswar

Punjab

40. Guru Angad Dev Veterinary and Animal Sciences University, Ludhiana
41. Punjab Agricultural University, Ludhiana

Rajasthan

42. MaharanaPratap University of Agriculture & Technology, Udaipur
43. Swami Keshwanand Rajasthan Agricultural University, Bikaner
44. Rajasthan University of Veterinary & Animal Sciences, Bikaner
45. SKN Agriculture University, Jobner
46. Agriculture University, Kota
47. Agriculture University, Jodhpur

Tamil Nadu

48. Tamil Nadu Agricultural University, Coimbatore
49. Tamil Nadu Veterinary & Animal Sciences University, Chennai
50. Tamil Nadu Fisheries University, Nagapattinam

Telangana

51. Sri Konda Laxman Telangana State Horticultural University, Hyderabad
52. Sri PV Narsimha Rao Telangana Veterinary University, Hyderabad
53. Professor Jayashankar Telangana State Agricultural University, Hyderabad

Uttarakhand

54. G.B. Pant University of Agriculture & Technology, Pantnagar
55. VCSG Uttarakhand University of Horticulture & Forestry, Bharsar

Uttar Pradesh

56. Chandra Shekhar Azad University of Agriculture & Technology, Kanpur
57. Narendra Deva University of Agriculture & Technology, Faizabad
58. SardarVallabhbhai Patel University of Agriculture & Technology, Meerut
59. U.P. Pt. Deen Dayal Upadhyaya Pashu Chikitsa Vigyan Vishwa Vidhyalaya Evem Go Anusandhan Sansthan, Mathura
60. Banda University of Agriculture and Technology, Banda
61. Sam Higginbottom University of Agriculture, Technology & Sciences, Allahabad

West Bengal

62. Bidhan Chandra Krishi Viswa Vidhyalaya, Mohanpur
63. West Bengal University of Animal & Fishery Sciences, Kolkata
64. Uttar Banga Krishi Viswavidhyalaya, Cooch Behar

Deemed Agricultural Universities

- Indian Agricultural Research Institute, New Delhi
- National Dairy Research Institute, Karnal
- Indian veterinary Research Institute, Izatnagar
- Central Institute on Fisheries Education, Mumbai